Mathcad for Electrical Engineers and Technologists

Mathcad tutorial with practical electrical examples

Stephen P. Tubbs, P.E.
formerly of the
Pennsylvania State University,
currently an
industrial consultant

NOTICE TO THE READER

The author does not warrant or guarantee any of the products, equipment, or software described herein or accept liability for any damages resulting from their use.

The reader is warned that electricity and the construction of electrical equipment are dangerous. It is the responsibility of the reader to use common sense and safe electrical and mechanical practices.

AMD Athlon XP is a trademark owned by American Micro Devices, Inc.

Intel Pentium 4 is a trademark owned by Intel Corporation.

Mathcad is a trademark owned by Parametric Technology Corporation (PTC).

Mathematica is a trademark owned by Wolfram Research, Inc.

MATLAB is a trademark owned by MathWorks, Inc.

Excel, *MSword*, *Vista*, *Windows*, and *XP Professional* are trademarks owned by the *Microsoft* Corporation.

Printed in the United States of America and United Kingdom.

ISBN   978-0-9819753-1-3

CONTENTS

iv

INTRODUCTION

Mathcad simultaneously solves and documents calculations. It is oriented toward non-programmers who need to solve numerical engineering problems. Users like Mathcad because its programs follow the natural format of manual calculations.

The main purpose of this book is to quickly teach an electrical engineer or technologist how to use Mathcad. Complete keystroke-to-keystroke details are provided for problem solution and documentation. The reader learns by example. The secondary purpose is to demonstrate Mathcad solutions of practical electrical problems.

As a calculating tool, Mathcad solves equations. The equations are entered into Mathcad in a format similar to that used in manual calculations. It will solve mesh equations with real or complex numbers and will solve differential equations. Outputs can be numerical or graphical. Mathcad will also do symbolic calculations, meaning that it can reduce complex systems of equations to simpler equations.

Documenting calculations is a major reason that Mathcad is used in modern industry. Calculations that in the past might have been recorded in notebooks, or even on easily lost scraps of paper, are now done with Mathcad to take advantage of the accuracy, neatness, traceability, and standardization it provides.

The simplest and most basic uses of Mathcad are in the first example problems. Later examples demonstrate more complex Mathcad capabilities. The reader could use the examples' solutions as models for his own Mathcad programs.

It is assumed that the reader has an analytical electrical background of the sort that would be gained in a university electrical engineering or electrical engineering technology program.

Mathcad is available in a free 30 day demonstration version. The key features of Mathcad can be learned in 30 days.

Stephen Tubbs, P.E.

1.0 WHAT IS MATHCAD?

Mathcad was released in 1986 as desktop software for the simultaneous solving and documenting of engineering and scientific calculations. It can be thought of as an "engineering notebook-style calculating program." Mathcad automatically evaluates equations as they are entered.

Mathcad was originally written by Allen Razdow of the Massachusetts Institute of Technology. Razdow was cofounder of the Mathsoft company. Mathcad's first version ran on the MS-DOS operating system. It was the first to have live editing of mathematical equations. Mathcad was also the first to automatically check engineering units.

In 2006 Parametric Technology Corporation (PTC) bought the Mathsoft company and took control of Mathcad. PTC also owns Pro/Engineer, an integrated 3-D CAD/CAM/CAE program. Pro/Engineer and Mathcad can seamlessly share data.

This book was written using Mathcad Version 14.0.

2.0 MATHCAD AND POPULAR ALTERNATIVE PROGRAMS

2.1 MATHCAD

All versions can be found through the Parametric Technology Corporation (PTC) website, http://www.ptc.com. See the website for current prices and sales promotions.

2.1.1 MATHCAD 30-DAY FREE TRIAL DOWNLOAD

The trial version is the same as the full version of Mathcad except that it will expire 30 days after it is installed. An account is needed with ptc.com. Register with http://www.ptc.com/appserver/common/account/basic.jsp.

Specific instructions on the downloading, installation, and start-up of the Mathcad program can be found on the PTC website.

2.1.2 MATHCAD STUDENT EDITION

This version is only for students, faculty, and staff for their academic use. It not intended for others. Current academic ID is required for purchase. At the time of publication, the list price of the program from PTC was $59.99.

It is also possible to get an "educational version" with the purchase of Brent Maxfield's book, (see reference 5, page 97). Maxfield's book comes with a companion CD-ROM containing a full non-expiring, educational purposes only, North America only, version of Mathcad. The list price for the book is $49.95.

2.1.3 MATHCAD SINGLE-USER VERSION

This is similar to the Student Edition. However, this version is also able to seamlessly integrate with software, including: ANSYS, AutoCAD, Bentley Microstation, CATIA, ESRD StressCheck 7, Excel, LabView, Pro/ENGINEER, Pro/INTRALINK, Solidworks, and Windchill.

Of course, this version can be used for commercial purposes. However, PTC states that the single-user license is not suitable for U.S. Government, Military, or large corporate settings.

At the time of publication, the list price for Mathcad was $1,195.00. However, PTC was then offering a special reduced price of $895.00 for Mathcad, a maintenance contract, and add-on modules. Use of a maintenance contract assures that all known software bugs have been removed from the program. Contact your PTC representative to determine current prices and which deal is the best for you.

2.1.4 MULTIPLE-USER VERSIONS

Mathcad can also be purchased in versions suitable for multiple users. It is sold in corporate and educational node-locked licenses and floating licenses. Contact your Mathcad sales representative for details on these.

2.2 MATHCAD TRAINING AND TUTORIALS

The Mathcad "Getting Started Primers" can be found by left-clicking on "Help" in the main menu and then on "Tutorials". The "Primers" are well written and .should be done by new Mathcad users.

PTC offers 2 days of classroom and virtual classroom training in "Mathcad Essentials" courses. These courses cover the essentials of Mathcad including its whiteboard interface and math toolbars. The classroom training is taught in various locations around the world. See the PTC website for times and locations. At the time of publication, the cost of the course in the U.S. is $900.

PTC also has one day classes titled, "Using Advanced Programming Techniques", "Using Advanced Plotting Techniques", "Configuring Applications Controls", and "Using Advanced Formatting Techniques". See the PTC website for details.

2.3 HOW POPULAR IS MATHCAD?

Mathcad brochures state:

1) "Mathcad is the world's most widely used engineering calculation tool."

2) "More than 250,000 professionals worldwide are using Mathcad to perform, document, manage and share calculation and design work."

3) Mathcad is used from "…aerospace to automotive to pharmaceuticals and beyond."

Mathcad is a popular program. At the time of writing, Monster.com listed 89 job advertisements that mentioned Mathcad. Furthermore, it is likely that many companies are presuming that new graduates have a background in Mathcad or a similar program.

2.4 HOW DOES MATHCAD COMPARE WITH MATLAB AND MATHEMATICA?

Mathcad is like a very sophisticated calculator. It does its best on notebook or scratch-paper type calculations. As a "sophisticated calculator", it is considered to be more user-friendly and intuitive than MATLAB and Mathematica. However, some consider Mathcad cumbersome to program. The list price of Mathcad's professional version is much less than that of MATLAB and Mathematica.

MATLAB is a program by MathWorks, Inc. MAT is short for **mat**rix and LAB is short for **lab**oratory. It is especially good at manipulating numerical data in matrices and arrays. MathWorks and other companies make many specialized add-on programs that use MATLAB as their operating environment. One of The MathWorks' commonly used add-ons is Simulink. It creates a diagram from standard and customizable function blocks to make it easier to visualize the problem being solved. MATLAB is generally recognized as better than Mathcad for sophisticated programming and complex graphics.

Mathematica is a program by Wolfram Research, Inc. Its founder stated that "Mathematica is a system for doing mathematics by computer." It was written with the mathematician and scientist in mind. It is good at symbolic calculations. Mathematica is also capable of doing electrical calculations and programming. It is often considered to be between Mathcad and MATLAB in sophisticated programming and graphics capabilities.

3.0 EXAMPLE PROBLEMS

Mathcad has a great number of features, functions, and methods of accessing those features and functions. It has many more than are needed by most electrical engineers and technologists. Many Mathcad functions can be accessed through toolbars, pull-down menus, and key strokes. In the following examples, usually only one method of accessing a function will be described. Furthermore, in the examples, only the required Mathcad features and functions are used. This is to allow the beginner to make use of Mathcad more quickly without being bogged down by unneeded information. Later, as one spends more time with Mathcad, learning more Mathcad details is warranted.

The order of the examples goes roughly from the simplest and most basic to the more complicated and probably less used. Every beginner should do the first couple of examples. Later examples can be studied as needed.

Each example contains a complete program. The body of this book is in 12 pt. Times New Roman font. Mathcad material is in the 12 pt. Arial font on a grey background.

3.1 STEADY-STATE ELECTRICAL CIRCUITS

3.1.1 SIMPLE DC CIRCUIT

Problem:

Solve for V1, V2, I, and the power to resistor R2 for the circuit of Figure 3-1-1-1. The following values are given: VS = 10 volts, R1 = 3 Ω, and R2 = 3 Ω.

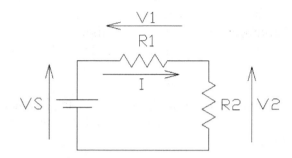

Figure 3-1-1-1 DC supply with a voltage divider.

Solution:

1) Start Mathcad. The blank screen seen in Figure 3-1-1-2 should appear. The "Mathcad Tips" window may appear with the worksheet.

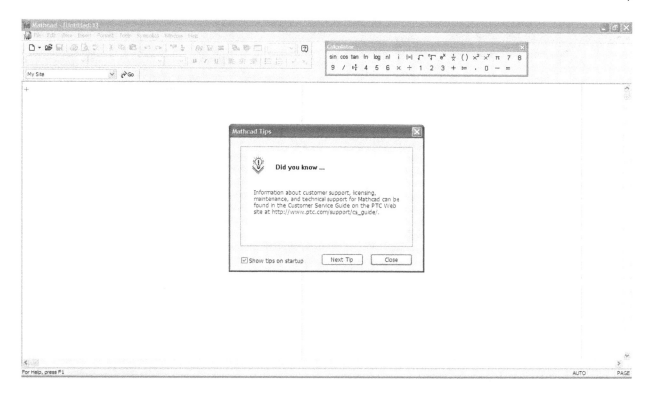

Figure 3-1-1-2 Blank Mathcad worksheet.

2) Close the "Mathcad Tips" window. Select the font to be "Normal", "Arial", and 12 pt. Sometimes, it is necessary to again select that font during the course of entering material into Mathcad. Mathcad may try to default to another font.

3) Left-click the cursor near the upper left of the blank worksheet and type in *VS:*. The result is in Figure 3-1-1-3. For easier viewing in this book, Mathcad text and calculations on the worksheet will be copied and entered directly into the text of this book. They will be put on a grey background to distinguish them.

$$VS := \blacksquare$$

Figure 3-1-1-3 Result of typing "VS:".

4) Notice how the ":" changed to a ":=" and a small filled rectangle appeared to the right of it. The small filled rectangle is a placeholder. When the placeholder is left-clicked, it is possible to put numbers or functions in its place.

5) Finish defining the variable "VS" by typing "10V". This defines it as 10 volts. The result is shown in Figure 3-1-1-4.

8

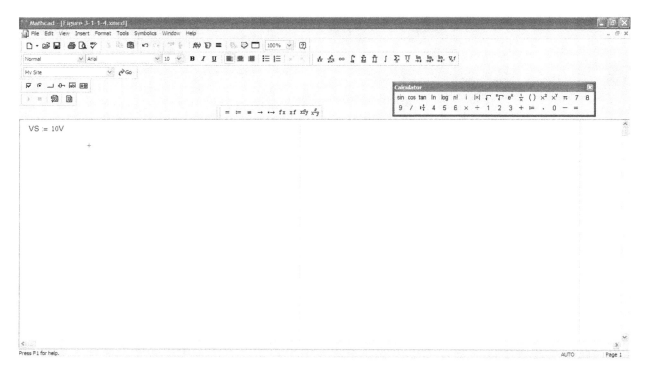

Figure 3-1-1-4 Mathcad worksheet after typing in "VS : 10 V".

6) Save the program in the directory of your choice as "Figure 3-1-1-1". Mathcad will automatically put its ".xmcd" extension after "Figure 3-1-1-1". As with all computer programs, save often.

7) Now, enter the following:

R1:3ohm
R2:3ohm
I:VS/R1+R2
V1:I*R1
V2:I*R2
P:V2*I
I=
V1=
V2=
P=

8) Mathcad simultaneously accepts the inputs and produces the outputs. The ":"s are used for data entry, The "="s are used to ask for values of variables. See the resulting program in Figure 3-1-1-5.

$$VS := 10\,V$$

$$R1 := 3\,ohm$$

$$R2 := 3\,ohm$$

$$I := \frac{VS}{(R1 + R2)}$$

$$V1 := I \cdot R1$$

$$V2 := I \cdot R2$$

$$P := V2 \cdot I$$

$$I = 1.667\,A$$

$$V1 = 5\,V$$

$$V2 = 5\,V$$

$$P = 8.333\,W$$

Figure 3-1-1-5 Mathcad SIMPLE DC CIRCUIT data input, equations and solutions.

9) Mathcad allows the blocks that make up the data input, equations, and solutions to be moved to different locations. To move "R2 := 3ohm" left-click on it. A rectangle should appear around it. Then, put the cursor on one of the rectangle edges. A black hand icon should appear. Left-click and hold down the button. Now, the rectangle can be dragged. Drag the blocks of Figure 3-1-1-5 to produce Figure 3-1-1-6. These blocks were aligned by simply using a straight edge held on the monitor screen.

$$VS := 10\,V \qquad R1 := 3\,ohm \qquad R2 := 3\,ohm$$

$$I := \dfrac{VS}{R1 + R2} \qquad V1 := I \cdot R1 \qquad V2 := I \cdot R2 \qquad P := V2 \cdot I$$

$$I = 1.667\,A \qquad V1 = 5\,V \qquad V2 = 5\,V \qquad P = 8.333\,W$$

Figure 3-1-1-6 Mathcad SIMPLE DC CIRCUIT data input, equations and solutions with blocks moved.

10) When Mathcad solves equations, it goes from left to right down a line and then left to right again in a zigzag fashion. The evaluation order was the same in Figure 3-1-1-5 as in Figure 3-1-1-6, so the results are the same.

11) Transpose the equations for I and V1. Now, Mathcad does not know what the value of I is when it tries to solve for V1. In the V1 equation the unknown I turns red (printed grey in this book) and the in solution line the value for V1 is not given. See Figure 3-1-1-7.

$$VS := 10\,V \qquad R1 := 3\,ohm \qquad R2 := 3\,ohm$$

$$V1 := I \cdot R1 \qquad I := \dfrac{VS}{R1 + R2} \qquad V2 := I \cdot R2 \qquad P := V2 \cdot I$$

$$I = 1.667\,A \qquad V1 = \blacksquare \qquad V2 = 5\,V \qquad P = 8.333\,W$$

Figure 3-1-1-7 Mathcad SIMPLE DC CIRCUIT data input, equations and solutions with the equations out of order.

12) Individual Mathcad blocks or groups of blocks can be dragged by another method. Select a block or group of blocks by left-clicking and holding the mouse button down when the cursor is at a corner of the block(s). Then, stretch a dashed line rectangle around them. This will cause the selected block(s) to have dashed line rectangles around them. Release the left mouse button. The dashed lines should still surround the selected block(s). Now, put the cursor on one of the rectangles. As before, a black hand will appear. Now, the selected blocks can be dragged to wherever desired.

13) The program in Figure 3-1-1-6 can be clarified by adding text. Make spaces for a text by dragging down blocks. The work sheet with spaces is shown in Figure 3-1-1-8.

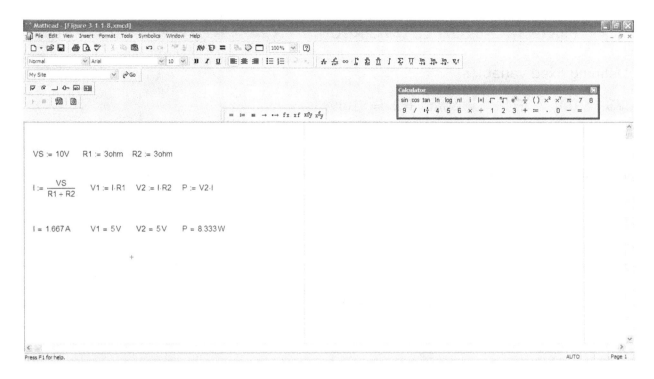

Figure 3-1-1-8 Mathcad SIMPLE DC CIRCUIT data input, equations and solutions with spaces created for text insertion.

14) To enter text, first place the cursor where you would like the text to begin and then left-click. This puts a red cross on the worksheet. Next, go to the main menu, select "Insert", and in that, select "Text Region". This will change the red cross to a small rectangle. Now, text can be typed. After text is entered, it can be moved around the sheet in blocks just as was done with the data, equations, and solutions. If it is necessary to change text after it has been entered, simply left-click the cursor on the old text and type in the new, just as in word processing. If the new text does not fit in the old text rectangle, Mathcad will word wrap it. Alternatively, the text rectangle can be stretched out by putting the cursor on the rectangle and, when a double pointed arrow appears, dragging the rectangle out. Notice that the file name, Figure 3-1-1-9.xmcd, is in the title. The result is shown in Figure 3-1-1-9.

12

Figure 3-1-1-9 SIMPLE DC CIRCUIT

Defining fixed variables:

$VS := 10V$ $R1 := 3\,ohm$ $R2 := 3\,ohm$

Equations:

$I := \dfrac{VS}{R1 + R2}$ $V1 := I \cdot R1$ $V2 := I \cdot R2$ $P := V2 \cdot I$

Results:

$I = 1.667\,A$ $V1 = 5\,V$ $V2 = 5\,V$ $P = 8.333\,W$

Figure 3-1-1-9   Mathcad SIMPLE DC CIRCUIT data input, equations and solutions with descriptive text.

3.1.2 DC MESH CIRCUIT

Problem:

Solve for the currents I1, I2, and I3 in the circuit of Figure 3-1-2-1. The following values are given: VS = 10 volts, R1 = 1 Ω, R2 = 2 Ω, R3 = 3 Ω, R4 = 4 Ω, R5 = 5 Ω, and R6 = 6 Ω.

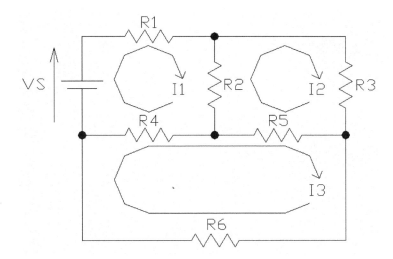

Figure 3-1-2-1 DC mesh circuit.

The simultaneous equations that describe Figure 3-1-2-1 are:

$$VS = I1 \cdot (R1 + R2 + R4) + I2 \cdot (-R2) + I3 \cdot (-R4)$$
$$0 = I1 \cdot (-R2) + I2 \cdot (R2 + R3 + R5) + I3 \cdot (-R5)$$
$$0 = I1 \cdot (-R4) + I2 \cdot (-R5) + I3 \cdot (R4 + R5 + R6)$$

Mathcad will be used to solve these equations three different ways. First, determinants and Cramer's rule are used to solve the equations. Second, a numerical approximation technique is used to solve the equations. Third, the equations are rewritten as a matrix equation and then solved by a numerical approximation technique.

14

3.1.2.1 DC Mesh Circuit Solution using Determinants

Mathcad's determinant solving ability is demonstrated in this example.

1) Mathcad uses a similar format to that used in Section 3.1.1. However, here determinants are used with Cramer's rule. The program is in Figure 3-1-2-1-1.

Figure 3-1-2-1-1.xmcd DC MESH ANALYSIS WITH DETERMINANTS

Defining fixed variables:

$VS := 10$ $R1 := 1$ $R2 := 2$ $R3 := 3$

$R4 := 4$ $R5 := 5$ $R6 := 6$

Equations:

$$D := \begin{vmatrix} R1 + R2 + R4 & -R2 & -R4 \\ -R2 & R2 + R3 + R5 & -R5 \\ -R4 & -R5 & R4 + R5 + R6 \end{vmatrix}$$

$$I1 := \frac{\begin{vmatrix} VS & -R2 & -R4 \\ 0 & R2 + R3 + R5 & -R5 \\ 0 & -R5 & R4 + R5 + R6 \end{vmatrix}}{D}$$

$$I2 := \frac{\begin{vmatrix} R1 + R2 + R4 & VS & -R4 \\ -R2 & 0 & -R5 \\ -R4 & 0 & R4 + R5 + R6 \end{vmatrix}}{D}$$

Continued

$$I3 := \frac{\begin{vmatrix} R1 + R2 + R4 & -R2 & VS \\ -R2 & R2 + R3 + R5 & 0 \\ -R4 & -R5 & 0 \end{vmatrix}}{D}$$

Results:

$I1 = 2.174$

$I2 = 0.87$

$I3 = 0.87$

Check:

$[I1 \cdot (R1 + R2 + R4) + I2 \cdot (-R2) + I3 \cdot (-R4)] = 10$

$[I1 \cdot (-R2) + I2 \cdot (R2 + R3 + R5) + I3 \cdot (-R5)] = 0$

$[I1 \cdot (-R4) + I2 \cdot (-R5) + I3 \cdot (R4 + R5 + R6)] = -1.776 \times 10^{-15}$

Figure 3-1-2-1-1 Mathcad DC MESH ANALYSIS WITH DETERMINANTS with input and output data, determinant equations, and check equations.

2) Notice that input voltage, resistances, and output currents do not have units. The Mathcad determinant operator requires that values be dimensionless. Mathcad's automatic unit checking cannot be used here. If units were required for report clarity, they could be added to the input and output data in "Text Region"s.

3) To make the determinant equation for *D*. Type *D:*. Open the "Math" toolbar and in that, open the "Vector and Matrix" toolbar. While the *D* placeholder is highlighted, left-click on the "Vector and Matrix" toolbar "Determinant" function. Then, left-click on the "Matrix or Vector" icon in the "Vector and Matrix" toolbar. Select a 3 by 3 matrix. At each placeholder, type in the variables shown in Figure 3-1-2-1-1.

4) Once the equation for *D* is made, it can be copied and then modified to create the numerators for the equations for "I1", "I2", and "I3". Note that there is a green zigzag line under "I1". "I1" is a built-in Mathcad "first order Bessel function of the first kind". The zigzag line is a warning that "I1" has been redefined. If the Mathcad "first order Bessel function of the first kind" were needed, "I1" should not be used for current.

16

3.1.2.2 DC Mesh Circuit Solution using Numerical Methods and Simultaneous Equations

Mathcad's "Solve Block" feature is demonstrated in this example.

1) Here Mathcad uses a numerical technique to find the solution to the equations. In the Mathcad program this numerical technique is called up with a "Solve Block". The program is in Figure 3-1-2-2-1.

Figure 3-1-2-2-1.xmcd DC MESH ANALYSIS WITH SIMULTANEOUS EQUATIONS

Defining fixed variables:

$VS := 10V$ $R1 := 1\,ohm$ $R2 := 2\,ohm$ $R3 := 3\,ohm$

$R4 := 4\,ohm$ $R5 := 5\,ohm$ $R6 := 6\,ohm$

Guess values:

$I1 := 1A$ $I2 := 1A$ $I3 := 1A$

Equations:

Given

$$VS = I1 \cdot (R1 + R2 + R4) + I2 \cdot (-R2) + I3 \cdot (-R4)$$

$$0V = I1 \cdot (-R2) + I2 \cdot (R2 + R3 + R5) + I3 \cdot (-R5)$$

$$0V = I1 \cdot (-R4) + I2 \cdot (-R5) + I3 \cdot (R4 + R5 + R6)$$

<Note "Ctrl="used after VS, 0V, and 0V>

$$\begin{pmatrix} I1val \\ I2val \\ I3val \end{pmatrix} := Find(I1, I2, I3)$$

Continued

Results:

I1val = 2.174 A

I2val = 0.87 A

I3val = 0.87 A

Check:

[I1val·(R1 + R2 + R4) + I2val·(−R2) + I3val·(−R4)] = 10 V

[I1val·(−R2) + I2val·(R2 + R3 + R5) + I3val·(−R5)] = 0 V

[I1val·(−R4) + I2val·(−R5) + I3val·(R4 + R5 + R6)] = 0 V

Figure 3-1-2-2-1 Mathcad DC MESH CIRCUIT WITH SIMULTANEOUS EQUATIONS with input and output data, equations, and solutions.

3) The "Align Regions" feature was used to make the calculations easier to read. To use it, drag a box around the text and math regions you want to align. Then, in the "Format" menu select "Align Regions" and choose "Across" or "Down". It is possible to align regions so that they appear on top of each other. If this happens, use the "Undo" feature in the "Edit" Menu.

4) Note that the Figure 3-1-2-2-1 program contains the following necessary "Solve Block" lines:

a) Initial values are provided above or to the left of the solved equations. These values give the Mathcad points from which to start its numerical analysis. In Figure 3-1-2-2-1, these are just below "**Guess values:**". They are:

I1 := 1 A   I2 := 1 A   I3 := 1 A

b) The "Given" statement is to be provided above or to the left of the equations. The "**Given**" statement is entered as a "Math Region" not a "Text Region".

c) The equals sign for the equations is special. Rather than simply typing "=", type "'Ctrl'=". On the Mathcad worksheet a **bold equals sign** appears. In Figure 3-1-2-2-1, these are the equal signs in the following equations:

$$VS = I1 \cdot (R1 + R2 + R4) + I2 \cdot (-R2) + I3 \cdot (-R4)$$
$$0V = I1 \cdot (-R2) + I2 \cdot (R2 + R3 + R5) + I3 \cdot (-R5)$$
$$0V = I1 \cdot (-R4) + I2 \cdot (-R5) + I3 \cdot (R4 + R5 + R6)$$

d) The "Find" statement in the next line tells Mathcad what values to solve for and where to record them. To create this statement, select the "Insert" menu. Then, select "Matrix". On this, choose one column and as many rows as variables being solved for. Type in where the solved values are to be stored in the matrix. In the parentheses after "Find" type in the corresponding variables. In Figure 3-1-2-2-1, the variables I1, I2, and I3 are solved for and their respective values stored in I1val, I2val, and I3val. This is shown below.

$$\begin{pmatrix} I1val \\ I2val \\ I3val \end{pmatrix} := Find(I1, I2, I3)$$

e) Equation solutions should be checked whenever the "Solve Block" is used. "Solve Block" solutions are approximations. Occasionally a "Solve Block" will produce incorrect results.

3.1.2.3 DC Mesh Circuit Solution using Numerical Methods and a Matrix Equation

Mathcad's matrix equation solving ability is demonstrated in this example.

Figure 3-1-2-3-1.xmcd DC MESH ANALYSIS
WITH A MATRIX EQUATION

Defining fixed variables:

$VS := 10V$ $R1 := 1\,ohm$ $R2 := 2\,ohm$ $R3 := 3\,ohm$

$R4 := 4\,ohm$ $R5 := 5\,ohm$ $R6 := 6\,ohm$

Guess values:

$I1 := 1A$ $I2 := 1A$ $I3 := 1A$

Equations:

Given

$$\begin{pmatrix} VS \\ 0V \\ 0V \end{pmatrix} = \begin{bmatrix} (R1+R2+R4) & -R2 & -R4 \\ -R2 & (R2+R3+R5) & -R5 \\ -R4 & -R5 & (R4+R5+R6) \end{bmatrix} \cdot \begin{pmatrix} I1 \\ I2 \\ I3 \end{pmatrix}$$

<Note "Ctrl=" used here>

$$\begin{pmatrix} I1val \\ I2val \\ I3val \end{pmatrix} := Find(I1,I2,I3)$$

Results:

$I1val = 2.174A$

$I2val = 0.87A$

$I3val = 0.87A$

Continued

20

Check:

$$
\begin{bmatrix}
(R1 + R2 + R4) & -R2 & -R4 \\
-R2 & (R2 + R3 + R5) & -R5 \\
-R4 & -R5 & (R4 + R5 + R6)
\end{bmatrix}
\cdot
\begin{pmatrix}
I1val \\
I2val \\
I3val
\end{pmatrix}
=
\begin{pmatrix}
10 \\
0 \\
0
\end{pmatrix} V
$$

Figure 3-1-2-3-1 Mathcad DC MESH CIRCUIT WITH A MATRIX EQUATION with input and output data, equations, and solutions.

1) The matrix equation was formed by left-clicking on "Insert" on the main menu and then on "Matrix". Select proper rows and columns, and then fill in the placeholders. As in Section 3-1-2-2, be certain to use a bold type "Ctrl =" rather than a simple "=".

2) The matrix equation for checking could be simply typed in again. An easier way would be to block the solution matrix equation, copy it with a "Ctrl c", and put the copy in place with a "Ctrl v". The equation copy would then be modified to be the "Check" equation. When modifying the "Check" matrix equation, place the cursor where you want to add or delete values, and then use the keyboard's "Space Bar" to move from one part of the equation to another. The use of the "Space Bar" is very important to the efficient entering and modifying of Mathcad equations. If you are not familiar with modifying equations and using the "Space Bar" in Mathcad, practice on these equations.

3.1.3 SIMPLE AC PHASOR CIRCUIT, NUMERICAL SOLUTION

Problem:

Determine the phasor values **V1** and **I** and the power to resistor R1 for the circuit of Figure 3-1-3-1. The following values are given: VS = 10 volts rms, XC1 = 3 Ω, and R1 = 3 Ω.

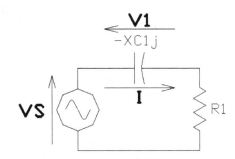

Figure 3-1-3-1 AC supply with a RC voltage divider.

Mathcad's "complex number", "absolute value/magnitude", "polar angle finding" features are demonstrated in this example.

Solution:

1) In this book's text and figures, phasor values will be indicated by larger bold print. For example, the phasor value for VS is **VS**. In the Mathcad program, it is not possible to change the font of individual characters. If a character is made bold and 14 point somewhere in a Mathcad program, then all the rest of the characters become bold and 14 point.

2) For calculations, Mathcad does not use the polar form of complex numbers. It only represents complex numbers in the Cartesian form, as a real part and an imaginary part. Extra program lines must be written to convert Mathcad Cartesian form complex numbers to the polar coordinate form used by phasors.

3) By default Mathcad uses an "i" to represent the square root of -1. It can use a "j", if "j" is selected. To select "j" left-click on "Format", "Results", and "Display Options". There "j" can be selected.

4) The "i" or "j" cannot be used alone to represent an imaginary number. The "i" or "j" must be preceded by a number or else Mathcad will think the "i" or "j" is a variable. For example, XCj would be entered as XC*1j. Later, after the cursor is off the equation where XC*1j was entered, Mathcad hides the 1. Then, the display will simply show XC*j.

5) Notice the use of the "arg( )" and "| |". The "arg( )" is the argument function. It finds the angle between -π radians and π radians of the complex number in its parentheses. "arg( )" is simply typed into the program. The "| |" is the absolute value function. It finds the absolute magnitude of the complex number it encloses.

6) The "Space Bar" is used here when composing the power equation. The "Space Bar" allows one to jump out of one function and start another. The key strokes for the program's power equation are:

        P:
        |x| <left-click on this in the "Calculator" toolbar>
        Space Bar
        ^
        2
        Space Bar
        *
        R1
        Enter

7) The data, program, and results are in Figure 3-1-3-2.

Figure 3-1-3-2.xmcd SIMPLE AC PHASOR CIRCUIT ANALYSIS, NUMERICAL SOLUTION

Defining fixed variables:

$$VS := 10\,V \qquad R1 := 3\,ohm \qquad XC1 := 3\,ohm$$

Equations:

$$I := \frac{VS}{\left[R1 + XC1 \cdot (-j) \right]}$$

$$V1 := XC1 \cdot I \cdot (-j)$$

$$magI := |I|$$
$$angleradI := arg(I)$$
$$angledegI := 57.2958\,arg(I)$$

$$magV1 := |V1|$$
$$angleradV1 := arg(V1)$$
$$angledegV1 := 57.2958\,arg(V1)$$

$$P := (|I|)^2 \cdot R1$$

Continued

24

Results:

$I = (1.667 + 1.667j) A$

$magI = 2.357 A$

$angleradI = 0.785$

$angledegI = 45$

$V1 = (5 - 5j) V$

$magV1 = 7.071 V$

$angleradV1 = -0.785$

$angledegV1 = -45$

$P = 16.667 W$

Figure 3-1-3-2 Mathcad AC PHASOR CIRCUIT with input data, equations, and numerical solution.

3.1.4 SIMPLE AC PHASOR CIRCUIT, SYMBOLIC SOLUTION

Problem:

Determine the phasor symbolic (equation) solutions for values **I** and **V1**, and the power to resistor R1 for the circuit of Figure 3-1-3-1.

Mathcad's symbolic equation solving ability is demonstrated in this example.

Solution:

1) Write the program of Figure 3-1-4-1. Notice that instead of an equals sign the "→" sign is used. Make this sign by typing "Ctrl." or by selecting it in the "Evaluation" toolbar.

Figure 3-1-4-1.xmcd SIMPLE AC PHASOR CIRCUIT ANALYSIS, SYMBOLIC SOLUTION

Equations:

$$I := \frac{VS}{\left[R1 + XC1 \cdot (-j) \right]}$$

$$V1 := XC1 \cdot I \cdot (-j)$$

$$magI := |I|$$
$$angleradI := arg(I)$$
$$angledegI := 57.2958 \, arg(I)$$

$$magV1 := |V1|$$
$$angleradV1 := arg(V1)$$
$$angledegV1 := 57.2958 \, arg(V1)$$

$$P := (|I|)^2 \cdot R1$$

Continued

26

Results:

$$I \rightarrow \frac{VS}{R1 - XC1 \cdot j}$$

$$magI \rightarrow \frac{|VS|}{|R1 - XC1 \cdot j|}$$

$$angleradI \rightarrow arg\left(\frac{VS}{R1 - XC1 \cdot j}\right)$$

$$angledegI \rightarrow 57.2958 \cdot arg\left[\frac{VS}{R1 - (1.0j) \cdot XC1}\right]$$

$$V1 \rightarrow -\frac{VS \cdot XC1 \cdot j}{R1 - XC1 \cdot j}$$

$$magV1 \rightarrow \frac{|VS \cdot XC1|}{|R1 - XC1 \cdot j|}$$

$$angleradV1 \rightarrow arg\left(-\frac{VS \cdot XC1 \cdot j}{R1 - XC1 \cdot j}\right)$$

$$angledegV1 \rightarrow 57.2958 \cdot arg\left[-\frac{(1.0j) \cdot VS \cdot XC1}{R1 - (1.0j) \cdot XC1}\right]$$

$$P \rightarrow \frac{R1 \cdot (|VS|)^2}{(|R1 - XC1 \cdot j|)^2}$$

Figure 3-1-4-1 Mathcad AC PHASOR CIRCUIT equations and symbolic solutions.

3.1.5 AC PHASOR MESH CIRCUIT

Problem:

Solve for the phasor currents **I1**, **I2**, and **I3** in the circuit of Figure 3-1-5-1. The following values are given: VS = 10 volts rms, R1 = 1 Ω, R2 = 4 Ω, XL1 = 2 Ω, XL2= 5 Ω, XC1 = 3 Ω, and XC2 = 6 Ω.

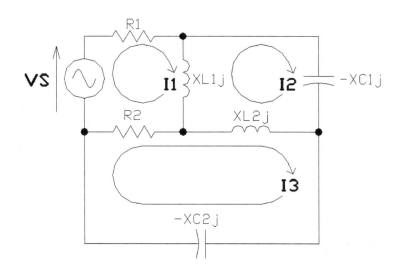

Figure 3-1-5-1 AC mesh circuit.

Solution:

1) The equations for this mesh circuit are:

$$\textbf{VS} = \textbf{I1}\cdot(R1 + R2 + XL1j) + \textbf{I2}\cdot(-XL1j) + \textbf{I3}\cdot(-R2)$$
$$0 = \textbf{I1}\cdot(-XL1j) + \textbf{I2}\cdot[(XL1 + XL2 - XC1)j] + \textbf{I3}\cdot(-XL2j)$$
$$0 = \textbf{I1}\cdot(-R2) + \textbf{I2}\cdot(-XL2j) + \textbf{I3}\cdot[R2 + (XL2 - XC2)j]$$

2) This circuit can be solved by the same methods as were used in the analysis of the DC mesh circuit, determinants, simultaneous equations, and matrix equations. The determinant method of solution is shown in Figure 3-1-5-2.

Figure 3-1-5-2.xmcd AC PHASOR MESH ANALYSIS WITH DETERMINANTS

Defining fixed variables:

$VS := 10$ $R1 := 1$ $R2 := 4$ $XL1 := 2$

$XL2 := 5$ $XC1 := 3$ $XC2 := 6$

Equations:

$$D := \begin{vmatrix} R1 + R2 + XL1 \cdot j & -XL1 \cdot j & -R2 \\ -XL1 \cdot j & (XL1 + XL2 - XC1) \cdot j & -XL2 \cdot j \\ -R2 & -XL2 \cdot j & R2 + (XL2 - XC2) \cdot j \end{vmatrix}$$

$$I1 := \frac{\begin{vmatrix} VS & -XL1 \cdot j & -R2 \\ 0 & (XL1 + XL2 - XC1) \cdot j & -XL2 \cdot j \\ 0 & -XL2 \cdot j & R2 + (XL2 - XC2) \cdot j \end{vmatrix}}{D}$$

$$I2 := \frac{\begin{vmatrix} R1 + R2 + XL1 \cdot j & VS & -R2 \\ -XL1 \cdot j & 0 & -XL2 \cdot j \\ -R2 & 0 & R2 + (XL2 - XC2) \cdot j \end{vmatrix}}{D}$$

$$I3 := \frac{\begin{vmatrix} R1 + R2 + XL1 \cdot j & -XL1 \cdot j & VS \\ -XL1 \cdot j & (XL1 + XL2 - XC1) \cdot j & 0 \\ -R2 & -XL2 \cdot j & 0 \end{vmatrix}}{D}$$

Continued

Results:

I1 = 1.478 + 0.27j

MagI1 := |I1| = 1.503   amps

AngleI1 := 57.2958 arg(I1) = 10.369   degrees

I2 = 0.489 + 1.176j

MagI2 := |I2| = 1.274   amps

AngleI2 := 57.2958 arg(I2) = 67.397   degrees

I3 = −0.2 + 0.832j

MagI3 := |I3| = 0.856   amps

AngleI3 := 57.2958 arg(I3) = 103.488   degrees

Check:

I1·(R1 + R2 + XL1·j) + I2·(−XL1·j) + I3·(−R2) = 10

I1·(−XL1·j) + I2·[(XL1 + XL2 − XC1)·j] + I3·(−XL2·j) = 0

I1·(−R2) + I2·(−XL2·j) + I3·[R2 + (XL2 − XC2)·j] = 0

Figure 3-1-5-2 Mathcad AC PHASOR MESH ANALYSIS CIRCUIT with input data, equations, and solutions.

3.1.6 AC INDUCTION MOTOR RECEIVING REDUCED VOLTAGE

Problem:

An AC three-phase 5 hp induction motor powers a fan. The power output of the motor is assumed to be a constant 5 hp. Using Mathcad and the motor's equivalent circuit, determine the motor's current, power loss, and slip with input voltages of 480 and 375 Vrms.

The values for the equivalent circuit of the motor are R1 = 1.6 Ω, R2 = 1.4 Ω, RM= 980 Ω, XM = 120 Ω, X = 6.5 Ω. These can be seen in Figure 3-1-6-1.

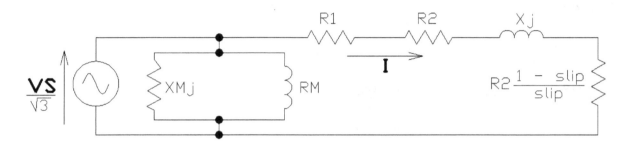

Figure 3-1-6-1 AC induction motor equivalent circuit of one phase.

Mathcad's "Square Root" feature is demonstrated in this example.

Solution:

1) Equations that describe the motor's output power, power loss, and stator/rotor circuit current are:

Motor output power (watts) $PO = 3 \cdot |\mathbf{I}|^2 \cdot R2 \cdot (1\text{-slip})/\text{slip}$

Rotor equivalent circuit current (amps) $\mathbf{I} = \mathbf{VS}/[\text{SQRT}(3)](R1 + R2/\text{slip} + Xj)$

Motor power loss (watts) $PL = 3 \cdot [|\mathbf{I}|^2 \cdot (R1 + R2) + |\mathbf{VS}/\text{SQRT}(3)|^2/RM]$

Total motor current (amps) $\mathbf{IT} = \mathbf{I} + \mathbf{VS}/\text{SQRT3})/RM + \mathbf{VS}/\text{SQRT3})/(XMj)$

2) The equations being solved are entered between the Mathcad "Solve Block's" "Given" statement and "Find" statement. The equation for PL was not involved in the numerical methods used in the "Solve Block", so it was written outside of the "Solve Block" and used an "**Ival**" rather than an "**I**" Note that the equations in the "Solve Block" require the use of the "Crtl=" equals sign rather than the ":=" or "=" equals signs. The program is in Figure 3-1-6-2.

3) To make a square root symbol in Mathcad type "\".

4) Notice how the output power of the motor is entered into Mathcad as 5 hp rather than 5 x 746 = 3730 watts. Mathcad automatically converts the 5 hp to watts, its standard units, for calculations.

Figure 3-1-6-2.xmcd INDUCTION MOTOR WITH REDUCED SUPPLY VOLTAGE

Defining fixed variables:

$VS := 480\,V$

$R1 := 1.6\,ohm$ $\qquad$ $R2 := 1.4\,ohm$ $\qquad$ $RM := 980\,ohm$

$XM := 120\,ohm$ $\qquad$ $X := 6.5\,ohm$ $\qquad$ $PO := 5\,hp$

Guess values:

$slip := .01$ $\qquad$ $I := (1 + j)A$

Equations:

Given

$$PO = 3\left(|I|\right)^2 \cdot R2 \cdot \frac{(1 - slip)}{slip}$$

<Note "Ctrl=" used after PO and VS>

$$VS = \sqrt{3} \cdot I \cdot \left(R1 + \frac{R2}{slip} + X \cdot j \right)$$

$$\binom{slipval}{Ival} := Find(slip, I)$$

Continued

32

$$PL := 3 \cdot \left[(|Ival|)^2 (R1 + R2) + \frac{\left(\frac{|VS|}{\sqrt{3}}\right)^2}{RM} \right]$$

$$IT := Ival + \frac{\frac{VS}{\sqrt{3}}}{RM} + \frac{\frac{VS}{\sqrt{3}}}{XM \cdot j}$$

Results:

slipval = 0.025

$Ival = (4.73 - 0.531j) \, A$

$|IT| = 5.762 \, A$

$PL = 0.439 \cdot kW$

Check:

$$3 (|Ival|)^2 \cdot R2 \cdot \frac{(1 - slipval)}{slipval} = 5 \cdot hp$$

$$\left| \sqrt{3} \cdot Ival \cdot \left(R1 + \frac{R2}{slipval} + X \cdot j \right) \right| = 480 \, V$$

Figure 3-1-6-2 Mathcad program solving for the current, power loss, and slip of an AC induction motor with 480 Vrms applied.

5) It is a good idea to have a "Check" section, whenever using a "Solve Block".

6) Notice the "hp" units for power in the first "Check" equation. Mathcad defaults to "W" <watts> units for power. In the program, it was converted by the following procedure.

a) The equation with the "W" is selected by left-clicking on it.

b) The "W" is selected by holding the left mouse button down and highlighting it. This causes an "Insert Unit" window to appear.

c) The appropriate unit ("hp" in this case) was selected and "OK" left-clicked in the window. The result is automatically converted to horsepower and the "hp" units appear after it.

7) The results when 375 V are applied are shown in Figure 3-1-6-3.

Results:

slipval = 0.045

Ival = $(6.314 - 1.243j)$ A

$|IT|$ = 7.211 A

PL = 0.516·kW

Figure 3-1-6-3 Results of the Mathcad program of Figure 3-1-6-2 with 375 Vrms applied.

3.1.7 AC INDUCTION MOTOR 2-D PLOT OF EFFICIENCY AND OUTPUT POWER VERSUS SLIP

Problem:

Use Mathcad to create a graph of %Efficiency versus Output Power for the motor of Section 3.1.6. Use an input voltage of 480 V rms and the same equivalent circuit as in Section 3.1.6.

Mathcad's "range variable", "function of a variable", and "2-D plotting with one and two vertical axes" features are demonstrated in this example.

Solution:

1) The equations needed are:

Rotor equivalent circuit current (amps) $\mathbf{I} = \mathbf{VS}/[\text{SQRT}(3)(R1 + R2/\text{slip} + Xj)]$

Motor output power PO (watts) $= 3 \cdot |\mathbf{I}|^2 \cdot R2 \cdot (1-\text{slip})/\text{slip}$

Motor power loss (watts) $PL = 3 \cdot [\,|\mathbf{VS}|^2/(3 \cdot RM) + |\mathbf{I}|^2 \cdot (R1+R2)]$

Efficiency (pu) $EFF = PO/(PO+PL)$

2) Enter the equations into Mathcad as shown in Figure 3-1-7-1.

Figure 3-1-7-1.xmcd INDUCTION MOTOR EFFICIENCY AND OUPUT POWER VERSUS SLIP

Defining fixed variables:

VS := 480 V

R1 := 1.6 ohm   R2 := 1.4 ohm   RM := 980 ohm

XM := 120 ohm   X := 6.5 ohm

Defining range variable:

slip := .001 , .002 .. .3

Continued

Equations:

$$I(slip) := \frac{VS}{\sqrt{3}\left(R1 + \frac{R2}{slip} + X \cdot j\right)}$$

$$PO(slip) := 3 \cdot \left(\left|I(slip)\right|\right)^2 \cdot R2 \cdot \frac{1 - slip}{slip}$$

$$PL(slip) := 3 \cdot \left[\frac{\left(\left|VS\right|\right)^2}{3 \cdot RM} + \left(\left|I(slip)\right|\right)^2 \cdot (R1 + R2)\right]$$

$$EFF(slip) := \frac{PO(slip)}{PO(slip) + PL(slip)}$$

Plots:

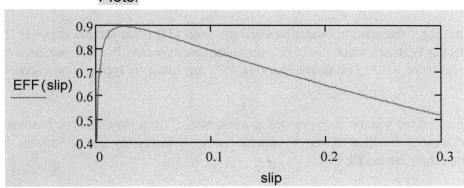

Continued

36

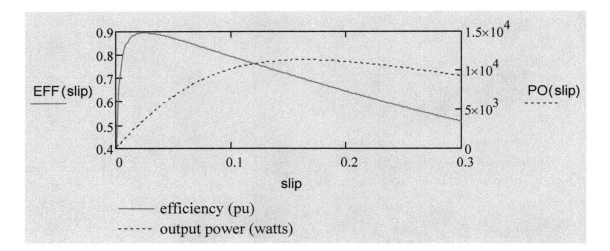

Figure 3-1-7-1 Mathcad program producing graphs of an AC induction motor's efficiency and output power versus slip.

3) The range and values of the x-axis variable used in plotting are defined by the statement:

$$slip := .001, .002 .. .3$$

The .001 is the first slip value used in evaluations and plotting. .002 is the second value. (.002 - .001) = .001 is the spacing between values of slip. Slip values incremented by .001 are used in evaluations until a final slip value of .3. The double periods, "..", are made by typing a semi-colon, ";".

4) Plotting requires that the y value be expressed as a function of the x variable. The Mathcad format needed for expressing a function is the y variable name followed by the x variable in parentheses. In the program the function for I is:

$$I(slip) := \frac{VS}{\sqrt{3}\left(R1 + \frac{R2}{slip} + X \cdot j\right)}$$

Functions with more than one variable can also be made in Mathcad.

5) The first plot shown is created with the following procedure:
      a) From the main menu select "Insert", "Graph", and then "X-Y Plot".
      b) A blank plot will appear.

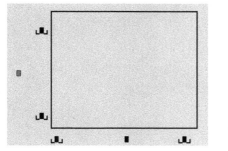

0

c) On the left center placeholder type *EFF(slip)*.
d) On the bottom center placeholder type *slip*.
e) The following will appear.

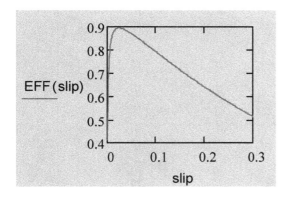

f) The plot can be stretched out by left-clicking in the center of it and then dragging it out by the placeholder-like rectangle at the bottom, right side, and right side bottom corner. The plot in Figure 3-1-7-1 was simply dragged out on the right side.

6) The second plot was started the same as the first, but then additions were made to it. The following additional steps created the modification.

a) Double left-click on the plot. The window in Figure 3-1-7-2 appears:

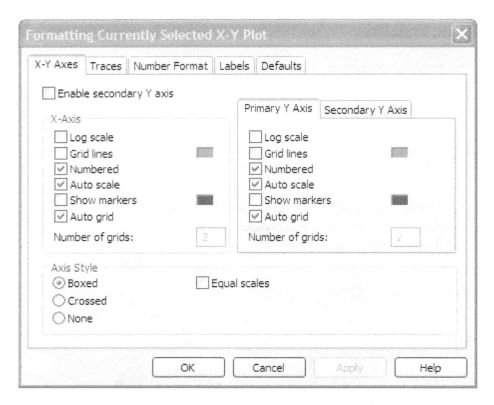

Figure 3-1-7-2 Plotting options window.

      b) Left-click on "Enable secondary Y-axis" and "OK".
      c) Type *PO(slip)* onto the right center placeholder. The second plot shown in Figure 3-1-7-1 appears.

3.1.8 AC INDUCTION MOTOR 3-D PLOT OF EFFICIENCY VERSUS SUPPLY
FREQUENCY AND SPEED

This uses Mathcad's "3-D plotting" feature. Mathcad can produce a 3-D plot of a function of two variables or of data. It can produce bar, contour, data point, patch, or vector field plots.

The plotted function can be a simple two-dimensional equation. An example would be $f(x,y) = x^2 + y^2$ where x represents the x-axis, y the y-axis, and $f(x,y)$ the z-axis. The plotted function can also be part of a system of equations that results in a z-axis value. An example would be $f(u,v) = u^2 + v$, $u = x$, and $v = y^2$. The example problem solved in this section uses a system of equations.

Mathcad is capable of producing a 3-D plot of a 2-D data matrix. For this, it would use the row and column location numbers as the x and y-axis plot values. Then, the data values at each location are used as the z plot values.

Mathcad is also capable of producing a 3-D parametric plot of the data of three identically sized 2-D data matrices where each data matrix represents one of the axes.

Mathcad cannot directly plot the results of systems of functions that require Mathcad "Solve Blocks" for solution.

Problem:
Use Mathcad to create a 3-D plot showing the efficiency of a motor producing a constant 5 hp output as its supply frequency and speed are varied. Use the same motor equivalent circuit and data as in Section 3.1.6.

Mathcad's "3-D plot" ability is demonstrated in this example.

Solution:
1) Definitions:

Motor output power (hp), POH

AC frequency (Hz), f

Motor speed (rpm), N

Motor equivalent resistances, see Figure 3-1-6-1 (Ω), R1, R2, and RM

2) The equations needed are:

Motor output power (watts), $PO = 746 \cdot POH$

Synchronous motor speed (4 pole motor) (rpm), $NSY = 30 \cdot f$

Motor slip (pu), $slip = 1 - N/NSY$

Line-to-line voltage (volts), $VS = 8 \cdot f$ <Note: To avoid magnetically saturating motor laminations, variable frequency drives often maintain a voltage to frequency ratio. A more thorough investigation would vary this ratio.>

Motor current (amps), $I = |\{PO(slip)/[3 \cdot R2 \cdot (1 - slip)]\}^{.5}|$

Motor power loss (watts), $PL = 3 \cdot [VS^2/(3 \cdot RM) + (R1 + R2) \cdot I^2]$

Efficiency (pu) $EFF = PO/(PO+PL)$

3) Enter the equations into Mathcad as shown in Figure 3-1-8-1.

4) To create the 3-D plot:
   a) Locate the cursor in the desired location.
   b) Type "CTRL2". This will make a blank 3 axis scatter plot appear.
   c) Type *EFF* and return onto the placeholder at the lower left of the plot.
   d) Double left-click on the plot. A "3-D Plot Format" window appears.
   e) Select "Axes". The "X-Axis" window appears by default.
   f) Left-click "Label" and type in *Frequency (Hz)*.
   g) Left-click on "Auto Scale" to turn it off.
   h) Type in a "Maximum Value" of *70* and a "Minimum Value" of *0*.
   i) Left-click on "Y-Axis". Enter *Speed (rpm)* as the "Label", *0* as the "Minimum Value" and *2000* as the "Maximum Value".
   j) Left-click on "Z-Axis". Enter *Efficiency (pu)* as the "Label", *0* as the "Minimum Value", and *1* as the "Maximum Value".
   k) Left-click on "Quick Plot Data".
   l) For "Range 1", "start" at *.001* and "end" at *70*. <*.001* is selected rather than *0* to avoid divide by zero errors>
   m) For "Range 2", "start" at *.001* and "end" at *2000*. <*.001* is selected rather than *0* to avoid divide by zero errors>
   n) Left-click on "Apply" and "OK"
   o) There are many adjustments that can be made to the plot. It can be turned by dragging its view. It can be increased or decreased in size. It can be changed to other plot types (bar, contour, data point, patch, or vector field plot). The learner should experiment with the plot options.

Figure 3-1-8-1.xmcd INDUCTION MOTOR EFFICIENCY 3-D PLOT

Defining fixed variables:

$R1 := 1.6$ $R2 := 1.4$ $RM := 980$ $POH := 5$

System of Equations:

$PO := 746 \cdot POH$

$NSY(f, N) := 30 \cdot f$

$slip(f, N) := 1 - \dfrac{N}{NSY(f, N)}$

$VS(f, N) := 8 \cdot f$

$I(f, N) := \left| \left[PO \cdot \dfrac{slip(f, N)}{3 \cdot R2 \cdot (1 - slip(f, N))} \right]^{.5} \right|$

$PL(f, N) := 3 \cdot \left[\dfrac{VS(f, N)^2}{3 \cdot RM} + (R1 + R2) \cdot I(f, N)^2 \right]$

$EFF(f, N) := \dfrac{PO}{PO + PL(f, N)}$

Continued

42

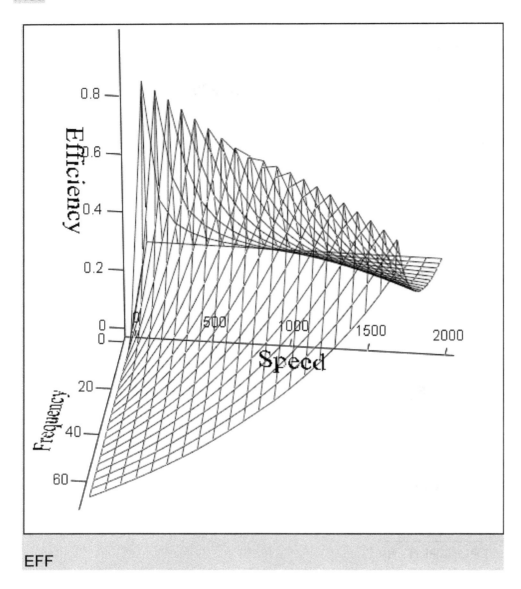

EFF

Figure 3-1-8-1 Mathcad program producing a 3-D plot of motor efficiency versus motor speed and input voltage frequency.

3.1.9 SIMPLE STATIC DC DIODE CIRCUIT

Problem:

Use Mathcad to determine the operating voltage and current of a diode in series with a DC voltage source and resistor. The circuit is shown in Figure 3-1-9-1. Use a resistor, R, of 5 Ω and a DC voltage supply, VS, of 1 volt. The diode characteristic curve data is supplied in a data chart.

Mathcad's "tabulated data", "plotting of data", "linear interpolation", "plotting of a function", and "Find" features are demonstrated in this example.

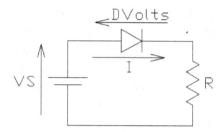

Figure 3-1-9-1 Diode in series with a DC supply and resistor.

Solution:

1) The voltage loop equation is:

$$VS = DVolts + I \cdot R$$

2) Create the program's data table by selecting "Insert" in the main menu and then selecting "Data" and "Table". See the program in Figure 3-1-9-2.

3) Type *Char* for the data table's name.

4) Enter data into the table by typing it in or blocking it, copying it, and pasting it from a program like Excel.
    a) To enter data by typing it in:
        1] Left-click on the data table so that small black rectangles appear around its border.
        2] Left-click on the lower center small black rectangle and drag it to lengthen the table to 13 rows.
        3] Enter the diode voltage and current characteristic values by typing them in.

b) To paste Excel spreadsheet data into a Mathcad table:
    1] Put the data into Excel.
    2] Block copy the Excel data.
    3] Make a data table in Mathcad.
    4] Right-click on the Mathcad table.
    5] Select "Paste Table".
    6] The Excel data is now in the Mathcad table.

5) To create the plot of the diode characteristic data point voltage versus current shown in the first plot in Figure 3-1-9-2.
    a) Initially, set up the plot as was done in Section 3-1-7.
    b) For the Y-axis enter "Char<Ctrl>61<Enter>". This will produce "Char$^{\langle 1 \rangle}$" on the Y-axis, corresponding to column 1 of the *Char* data table.
    c) For the X-axis enter "Char<Ctrl>60<Enter>". This will produce "Char$^{\langle 0 \rangle}$" on the X-axis, corresponding to column 0 of the *Char* data table.
    d) The X-axis limits are set to -20 and 10.
    e) The Y-axis limits are set to -40 and 40.
    f) To title the plot double left-click on the plot. Left-click on the "Labels" tab. On the "Title" type *Diode Current vs Voltage from data points*. Then, left-click "OK".

6) The linear interpolation function, "linterp", is used to convert the data points to a function that connects the data points with straight lines between the data points.

$$I(DVolts) := linterp\left(Char^{\langle 0 \rangle}, Char^{\langle 1 \rangle}, DVolts\right)$$

I(DVolts) is a function that is based on X data values of "Char$^{\langle 0 \rangle}$" and corresponding Y data values of "Char$^{\langle 1 \rangle}$". I(DVolts) corresponds to the Y-axis and DVolts corresponds to the X-axis.

7) The second plot is of the function I(DVolts) versus DVolts. Notice that it is practically the same as the data point plot and it has extrapolated values beyond the data points.

8) The "Find" function solves the voltage loop equation and the I(DVolts) function for diode voltage and current.

Figure 3-1-9-2.xmcd SIMPLE STATIC DC DIODE CIRCUIT

Tabulated diode voltage (volts) versus current
(amps):

Char :=

	0	1
0	-11.7	-30
1	-11	-2.5
2	-10.9	-1
3	-10	-0.2
4	-5	-0.05
5	0	0
6	0.7	0.1
7	0.75	0.2
8	0.85	0.5
9	1	1
10	1.2	2
11	1.5	8
12	2	20
13	2.2	30

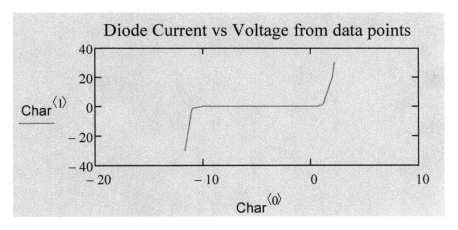

Diode Current vs Voltage from data points

$\overline{Char^{\langle 1 \rangle}}$

$Char^{\langle 0 \rangle}$

Continued

$$I(DVolts) := linterp\left(Char^{\langle 0 \rangle}, Char^{\langle 1 \rangle}, DVolts\right)$$

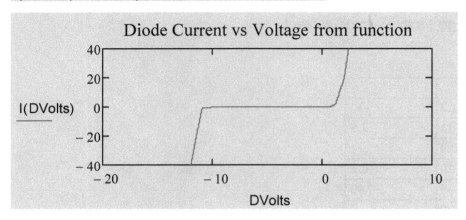

$R := 5$

$VS := 1$

Guess Values:

$DVolts := .1$

Given

$VS = DVolts + R \cdot I(DVolts)$ <Note "Ctrl=" after VS>

$DVoltsval := Find(DVolts)$

Results:

$DVoltsval = 0.583$

$I(DVoltsval) = 0.083$

Check:

$DVoltsval + R \cdot I(DVoltsval) = 1$

Figure 3-1-9-2 Mathcad program determining the voltage across and current through a diode.

3.1.10 SIMPLE STEADY-STATE AC DIODE CIRCUIT

Problem:

Use Mathcad to determine the voltage versus time across a diode in series with an AC voltage source and resistor. The circuit is shown in Figure 3-1-10-1. Use a resistor, R, equal to 5 Ω and an AC voltage supply, VS, equal to 20·sin(377·t). The diode data is the same as used in Section 3.1.9.

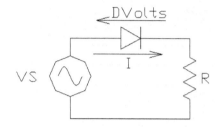

Figure 3-1-10-1 Diode in series with an AC supply and resistor.

Solution:

1) The voltage loop equation is:

$$VS = 20 \cdot \sin(377 \cdot t) = DVolts + I \cdot R$$

2) The diode characteristic table and linearly interpolated function are the same as in Section 3-1-9.

3) The "Find" function is similar to that used in Section 3-1-9. However, this time DVolts is a function of a varying VS rather than a fixed value (DVoltsval).

4) After the "Find" function VS is expressed as a function of time, t.

48

5) The output plot of DVolts versus time, t, shows that DVolts is a function of VS which is in turn a function of time, t.

Figure 3-1-10-2.xmcd SIMPLE STEADY-STATE AC DIODE CIRCUIT

Tabulated diode voltage (volts) versus current (amps):

Char :=

	0	1
0	-11.7	-30
1	-11	-2.5
2	-10.9	-1
3	-10	-0.2
4	-5	-0.05
5	0	0
6	0.7	0.1
7	0.75	0.2
8	0.85	0.5
9	1	1
10	1.2	2
11	1.5	8
12	2	20
13	2.2	30

$I(DVolts) := linterp\left(Char^{\langle 0 \rangle}, Char^{\langle 1 \rangle}, DVolts\right)$

$R := 5$

Continued

Guess Values:

DVolts := .1

Given

VS = DVolts + R·I(DVolts)   <Note "Ctrl=" after VS>

DVolts(VS) := Find(DVolts)

VS(t) := 20·sin(377·t)

Results:

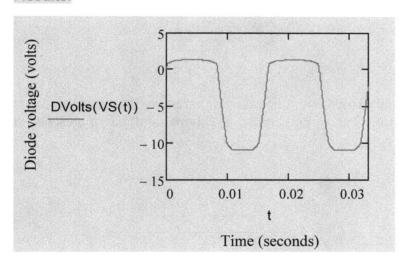

Figure 3-1-10-2 Mathcad program determining the AC voltage across a diode versus time.

3.1.11 INDUCTOR CHARACTERISTICS

Problem:

Flux versus current curves are known for an iron core inductor. Ignoring eddy current losses, what is the voltage across its coil and the power delivered to it when a sinusoidal AC current is forced through it? The sinusoidal current is 6 amps peak at 60 Hz. The magnetization curve, flux (Wb) versus current (amps.) is given in Figure 3-1-11-1. Assume there is no resistance in the circuit. The electrical circuit is shown in Figure 3-1-11-2.

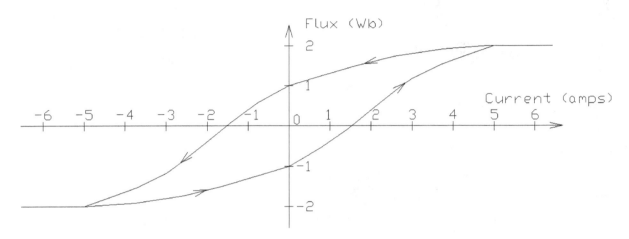

Figure 3-1-11-1 Inductor hysteresis curves. The vertical axis represents the flux enclosed by the coil (flux in the iron times the number of coil turns). The horizontal axis represents the current through the coil.

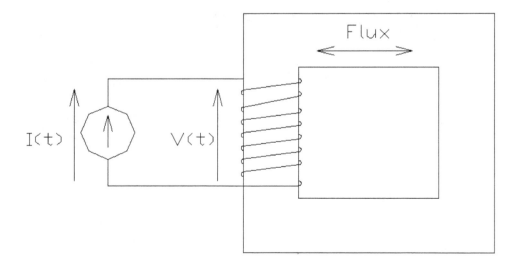

Figure 3-1-11-2 Inductor circuit. The supply is an ideal current source that forces sinusoidal 6 amp peak 60 Hz current through the inductor.

Mathcad's "Add Line operator", "if", "<" (less than Boolean statement), "≥" (greater than or equal to Boolean statement), "≤" (less than or equal to Boolean statements), "∧" (AND Boolean statement), and the "Calculate Worksheet command" are demonstrated in this example.

Solution:

1) The magnetization curves in Figure 3-1-11-1 apply to the condition where the current is steadily varying from less than -5 amps to greater than 5 amps. If a smaller current were applied, then curves with a lesser current range would be needed. This magnetization curve does not apply to the startup condition, where the magnetization curve would start at zero flux and zero current.

2) As already stated, this problem is only concerned with the hysteresis losses. A more complete analysis of the inductor would also consider its eddy current losses.

3) The magnetization curves of Figure 3-1-11-1 are simplified and represented by linear equations. There are equations for the upper and lower curves. Different linear equations are used in different current ranges.

The Mathcad program of Figure 3-1-11-3 has the magnetizing curve generating equations in "Add Line operators" along with conditional "if", "<","≥", and "≤" statements. "Add Line operators" allow Mathcad to be programmed with statements like those used in BASIC and other programming languages. The output of an "Add Line operator" is one scalar number. It will not produce a vector or matrix quantity. "Add Line operators" and their associated statements operate as functions. For example, the first one, "FU(I)", acts as a function of "I".

The "Add Line operator" and "if" statement is entered into the program using the "Programming" toolbar. The "if" statement can not be entered by typing "i" and "f" on the keyboard, Mathcad would not recognize it. The "<" and "≤" statements should be inserted using the "Boolean" toolbar.

52

Figure 3-1-11-2.xcmd INDUCTOR HYSTERESIS

Magnetizing Curve Generating Equations

$$FU(I) := \begin{cases} -2 & \text{if } I < 5 \\ .4 \cdot I & \text{if } -5 \leq I < -3 \\ 1 + 2.2 \cdot \dfrac{I}{3} & \text{if } -3 \leq I < 0 \\ 1 + .8 \cdot \dfrac{I}{3} & \text{if } 0 \leq I < 3 \\ 1.5 + .1 \cdot I & \text{if } 3 \leq I < 5 \\ 2 & \text{if } 5 \leq I \end{cases}$$

$$FL(I) := \begin{cases} -2 & \text{if } I < 5 \\ -1.5 + .1 \cdot I & \text{if } -5 \leq I < -3 \\ -1 + .8 \cdot \dfrac{I}{3} & \text{if } -3 \leq I < 0 \\ -1 + 2.2 \cdot \dfrac{I}{3} & \text{if } 0 \leq I < 3 \\ .4 \cdot I & \text{if } 3 \leq I < 5 \\ 2 & \text{if } 5 \leq I \end{cases}$$

Magnetizing Curve

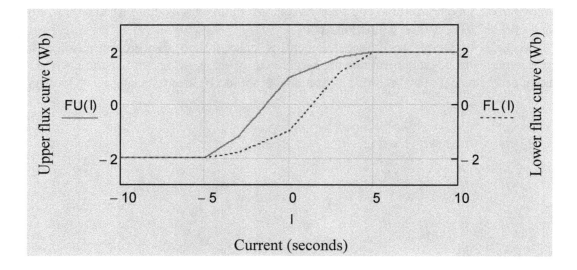

Continued

Determining Voltage and Power When AC Current is Applied

$$I(t) := 6 \cdot \sin(2 \cdot \pi \cdot f \cdot t) \qquad\qquad SLOPE(t) := \cos(2 \cdot \pi \cdot f \cdot t)$$

$$F(t) := \begin{cases} -2 & \text{if } I(t) < -5 \\[2mm] .4 \cdot I(t) & \text{if } -5 \le I(t) < -3 \wedge SLOPE(t) < 0 \\[2mm] 1 + 2.2 \cdot \dfrac{I(t)}{3} & \text{if } -3 \le I(t) < 0 \wedge SLOPE(t) < 0 \\[2mm] 1 + .8 \cdot \dfrac{I(t)}{3} & \text{if } 0 \le I(t) < 3 \wedge SLOPE(t) < 0 \\[2mm] 1.5 + .1 \cdot I(t) & \text{if } 3 \le I(t) < 5 \wedge SLOPE(t) < 0 \\[2mm] -1.5 + .1 \cdot I(t) & \text{if } -5 \le I(t) < -3 \wedge SLOPE(t) \ge 0 \\[2mm] -1 + .8 \cdot \dfrac{I(t)}{3} & \text{if } -3 \le I(t) < 0 \wedge SLOPE(t) \ge 0 \\[2mm] -1 + 2.2 \cdot \dfrac{I(t)}{3} & \text{if } 0 \le I(t) < 3 \wedge SLOPE(t) \ge 0 \\[2mm] .4 \cdot I(t) & \text{if } 3 \le I(t) < 5 \wedge SLOPE(t) \ge 0 \\[2mm] 2 & \text{if } 5 \le I(t) \end{cases}$$

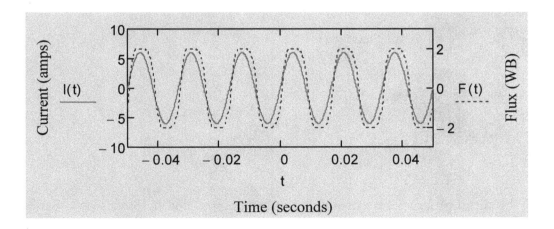

Continued

54

$$V(t) := \frac{d}{dt} F(t)$$

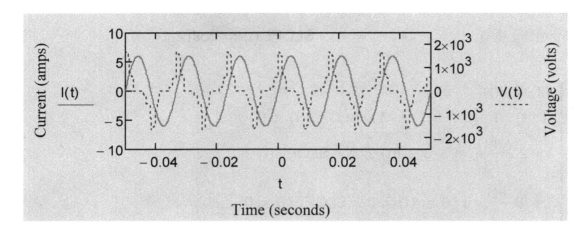

$$P(t) := V(t) \cdot I(t)$$

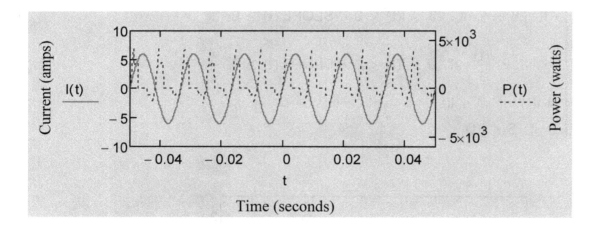

$$PRMS := f \cdot \int_0^{\frac{1}{f}} P(t) \, dt = 540$$

Figure 3-1-11-2 Mathcad program determining the voltage across and hysteresis power used by an iron core inductor receiving a sinusoidal current.

4) The upper and lower curve functions, "FU(I)" and "FL(I)" are plotted versus current. Plotting helps check that the curves have been properly represented by the "Add Line operator" programs.

5) The next program section determines the voltage across and power used by the inductor when AC sinusoidal current is forced through it. The frequency is set at 60 Hz and the current is set at 6·sin(2·π·f·t) amps. The equation SLOPE(t) = cos(2·π·f·t) is used to determine if upper or lower magnetization curve should be used. When SLOPE(t) is less than zero the current is decreasing and the upper magnetization curve is used. When SLOPE(t) is greater than or equal to zero the current is increasing and the lower magnetization curve is used.

6) The "Add Line operator" program determines the flux using the same conditional ranges as above, but in just one equation. This equation uses the Boolean AND statement, "∧". The "∧" determines if the upper or lower curve should be used.

7) Flux and current are plotted versus time. Note the flat tops on the flux waveform where the iron saturates.

8) The voltage across the inductor coil is the derivative of the flux with respect to time. That voltage is also plotted.

9) The hysteresis power used by the coil is the voltage across the coil times the current through it. That power is plotted.

10) The average hysteresis power is determined by an integral equation in the Mathcad program.

11) Note that the average hysteresis power determined in the program is 540 watts for a 60 Hz current supply. If the frequency, "f" in the program, were doubled to 120 Hz one would expect the power to double to 2x540 = 1080 watts. Replace the 60 in the program by 120 and left click the cursor near but outside of the "f := 120 area. The plots will correctly change, but Mathcad will not change the 540 watts, as it should.

To make Mathcad properly calculate all values in a worksheet use the "Calculate Worksheet" command. This command is in the "Tools" menu under "Calculate". When used, Mathcad will properly recalculate all values in its worksheet. In this example, it properly calculates 1.08×10^3 watts for a 120 Hz supply frequency.

3.2 TRANSIENT ELECTRICAL CIRCUITS

3.2.1 RL CIRCUIT, FIRST ORDER DIFFERENTIAL EQUATION

Problem:

Using Mathcad, produce a plot of current through the circuit of Figure 3-2-1-1 versus time. The following values are given: VS = 10 volts, R1 = 3 Ω, and L1 = 3 milliH. The switch closes at time t = 0 seconds and the initial current is 0 A.

Mathcad's " ′ (first order differentiation indicating symbol)" and "ordinary differential equation solving feature" are demonstrated in this example.

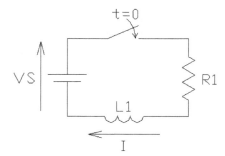

Figure 3-2-1-1 Simple inductor transient circuit.

Solution:

1) The loop equation for the circuit is:

$$VS(t) = R1{\cdot}I(t) + L1{\cdot}dI(t)/dt$$

The initial condition is I(0) = 0 A

Figure 3-2-1-2.xmcd DIFFERENTIAL EQUATION SOLUTION OF TRANSIENT RL CIRCUIT

$VS := 10$

$R1 := 3$

$L1 := .003$

Given

$R1 \cdot I(t) + L1 \cdot I'(t) = VS$ <Note "Ctrl=" after $R1 \cdot I(t) + L1 \cdot I'(t)$ and $I(0)$.>

$I(0) = 0$

$I := Odesolve(t, .005)$

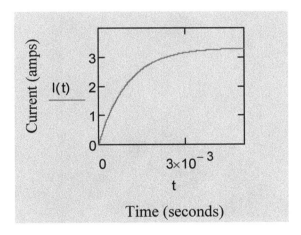

Figure 3-2-1-2 Mathcad program determining the current through a RL circuit.

2) Notice how the differential term "I'(t)" and the initial condition "I(0)" are to the left of their respective "Ctrl=" signs. This left-side location is a requirement of the "Odesolve" function.

3) "Odesolve" directs Mathcad to select what it estimates to be the best differential equation solving method for solving the equations between it and the "Given" statement. "I" is the function being solved for, the circuit current. In this case, it is a scalar. However, in other programs it could be a vector, see the program in Section 3.2.4. Only the minimum required parameters have been placed after the "Odesolve", "t, .005". The "t", time, is the variable for the function "I". The ".005" is maximum value of "t" for which the differential equation will be solved.

There is an optional number that could have been placed after the ".005". For discussion, this will be called "a". With "a" the "Odesolve" function would be written as "Odesolve(t, .005,"a")". "a" is the number of solution values that "Odesolve" solves for between the initial time, 0, and the end time, ".005". If "a" is not specified, "Odesolve" solves for a default 1000 solution values.

Odesolve chooses the equation solving method. However, if desired, it is possible to direct Mathcad to use a specific differential equation solving method. The specific methods will not be considered here. Information on the specific methods can be obtained from Mathcad "Help".

3.2.2 RLC CIRCUIT, SECOND ORDER DIFFERENTIAL EQUATION

Problem:

 Use Mathcad to produce a plot of the current through the circuit of Figure 3-2-2-1 versus time. The following values are given: VS = 10 volts, R1 = 3 Ω, L1 = 3 mH, and C1 = 3 microF. The switch closes at time t = 0 seconds, the initial current is 0 amps, and the initial voltage across the capacitor is 0 volts.

 Mathcad's " ″ (second order differentiation indicating symbol)″ is demonstrated in this example.

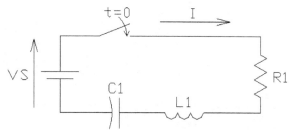

Figure 3-2-2-1 DC capacitor/inductor/resistor circuit.

Solution:

 1) The loop equation for the circuit is:

 VS = R1·I(t) + L1·dI(t)/dt + (1/C1)· $\int$ I(t)dt

 2) The equation is differentiated to put it in a form that Mathcad can solve.

 d/dt(VS) = d/dt[R1·I(t) + L1·dI(t)/dt + $\int$ [I(t)/C1]dt]

 This reduces to:

 0 = R1·dI(t)/dt + L1·d^{2}I(t)/dt^2+ I(t)/C

 The initial conditions are:

 I(0) = 0 A

 dI(0)/dt = VS/L1 <note this is at a time immediately after t = 0>

 3) These equations are converted to the Mathcad differential equation format. The program is shown in Figure 3-2-2-2.

Figure 3-2-2-2.xmcd DIFFERENTIAL EQUATION SOLUTION OF A TRANSIENT RLC CIRCUIT

$VS := 10$

$R1 := 3$

$L1 := .003$

$C1 := 3 \cdot 10^{-6}$

Given

$R1 \cdot I'(t) + L1 \cdot I''(t) + \dfrac{I(t)}{C1} = 0$

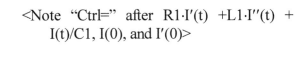

<Note "Ctrl=" after $R1 \cdot I'(t) + L1 \cdot I''(t) + I(t)/C1$, $I(0)$, and $I'(0)$>

$I(0) = 0$

$I'(0) = \dfrac{VS}{L1}$

$I := \text{Odesolve}(t, .005)$

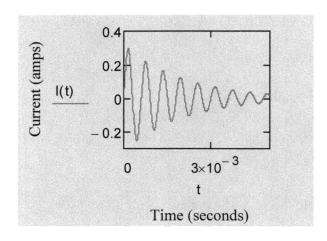

Figure 3-2-2-2 Mathcad program determining the current through a RLC circuit.

3.2.3 SIMPLE RL CIRCUIT, FIRST 0RDER DIFFERENTIAL EQUATION SYMBOLICALLY SOLVED WITH LAPLACE TRANSFORMS

Laplace transforms can provide analytical solutions to many linear constant coefficient differential equations. They were the usual method of time based differential equation solution before computers were available.

The steps to carry out a manual solution of a time-based linear constant coefficient differential equation with Laplace transforms are:

1) Write out the differential equation in its time-domain form and define its initial conditions.

2) Do a Laplace transformation of each side of the equation to create s-domain equations.

3) Insert initial condition values.

4) Algebraically solve the equation for the its s-domain function.

5) Using Laplace transform tables, find the inverse transformation of the s-based function.

Mathcad is able to Laplace transform multi-order linear constant coefficient time-domain differential equations to s-domain equations and inverse-transform s-domain equations back to time-domain equations.

The following example demonstrates Mathcad's Laplace transform ability on a simple problem. Mathcad is capable of doing Laplace transformation solutions of more complicated higher order differential equations.

Problem:

Using Mathcad's Laplace transform ability, produce the equation for and plot of the current through the circuit of Figure 3-2-1-1 versus time. The following values are given: VS = 10 volts, R1 = 3 Ω, and L1 = 3 milliH. The switch closes at time t = 0 seconds and the initial current is 0 A.

Mathcad's "d/dt differentiation operator", "laplace transform function", "inverse laplace transform function", and "simplify function" are demonstrated in this example.

Solution:

1) As before, the loop equation and initial condition for the circuit is:

VS(t) = R1·I(t) + L1·dI(t)/dt

I(0) = 0 A

2) These are entered into Mathcad using the d/dt operator. That operator is available on the Mathcad "Calculus" toolbar. Notice that this is different from the prime marks that were used in the programs of Sections 3.2.1 and 3.2.2.

3) The program is shown in Figure 3-2-3-1. For clarity, the program was written with explanatory comments in it.

Figure 3-2-3-1.xmcd LAPLACE TRANFORM SYMBOLIC SOLUTON OF A FIRST ORDER DIFFERENTIAL EQUATION

Enter right side and left side of the time domain equation.

$$Vrightside(t) := R1 \cdot I(t) + L1 \cdot \left(\frac{d}{dt} I(t) \right)$$

$$Vleftside(t) := VS$$

Use the "laplace" operator from the "Symbolic" toolbar on the right and left side equations.

$$Vrightside(t) \ laplace \ \rightarrow R1 \cdot laplace(I(t), t, s) - L1 \cdot I(0) + L1 \cdot s \cdot laplace(I(t), t, s)$$

$$Vleftside(t) \ laplace \ \rightarrow \frac{VS}{s}$$

Rewrite the equations, apply initial conditions and change laplace (I(t),t,s) to I(s). Block copy (Ctrlc), paste (Ctrlv), and Find & Replace commands would make this easier on larger equations.

$$Vrightside(s) := R1 \cdot I(s) + L1 \cdot s \cdot I(s)$$

$$Vleftside(s) := \frac{VS}{s}$$

Continued

Use the "Solve" function from the "Symbolic" toolbar to solve for I(s)

$$\text{Vrightside}(s) = \text{Vleftside}(s) \text{ solve}, I(s) \rightarrow \frac{VS}{s \cdot (R1 + L1 \cdot s)}$$

Define I(s) as the solution found above. Again, block, copy (Ctrlc) and paste (Ctrlv) will make this easier on larger equations.

$$I(s) := \frac{VS}{s \cdot (R1 + L1 \cdot s)}$$

Solve for the time-domain of I(s), I(t), using the inverse laplace,"invlaplace", function of the "Symbolic" toolbar.

$$I(s) \text{ invlaplace} \rightarrow -\frac{VS \cdot \left(e^{-\frac{R1 \cdot t}{L1}} - 1 \right)}{R1}$$

The solution is:

$$I(t) := -\frac{VS \cdot \left(e^{-\frac{R1 \cdot t}{L1}} - 1 \right)}{R1}$$

Continued

64

Checking that the equation for I(t) is a solution by substituting the equation for I(t) into it and simplifying it using the "simplify" function in the "Symbolic" toolbar.

$$R1 \cdot I(t) + L1 \cdot \left[\frac{d}{dt} - \frac{VS \cdot \left(e^{-\frac{R1 \cdot t}{L1}} - 1 \right)}{R1} \right] \text{ simplify } \rightarrow VS$$

The equation for I(t) can also be plotted to check that it is the same solution as was found in Section 3.2.1, Figure 3-2-1-2.

Values are assigned:

VS := 10   R1 := 3   L1 := .003

Define I(t) as the function found with inverse laplace above.

$$I(t) := -\frac{VS \cdot \left(e^{-\frac{R1 \cdot t}{L1}} - 1 \right)}{R1}$$

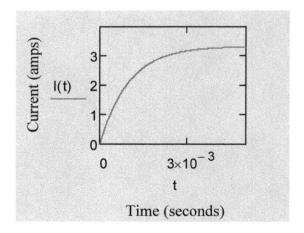

Figure 3-2-3-1 Mathcad program determining the equation through a RL circuit by Laplace Transforms.

3.2.4 AC SERIES MOTOR DRIVING A RECIPROCATING PUMP, SYSTEM OF FIRST ORDER DIFFERENTIAL EQUATIONS

Problem:

A 1/100 hp single-phase low speed series motor is directly driving a reciprocating pump. Use Mathcad to plot the current waveform, torque, and speed for .2 seconds after starting.

Equivalent circuit values are: R = 300 Ω, L = 1 H, VS = 120 Vrms, K4 = 10 watts/(amp^2·rpm), and f = 60 Hz.

The pump load torque can be modeled with the equation: T = K0 + K1·N(t) + K2·sin{(2·π·[N(t)/60]·t} + K3·dN/dt. Here K0 = .01 ft.lb., K1 = .00001 ft.lb./rpm, K2 = 1 ft.lb., and K3 = .0001 ft.lb.·sec/rpm.

The speed of the motor at time 0, N(0), is 0 rpm.

The initial current to the motor at time 0, I(0), is 0 amps.

The equivalent circuit of the series motor is in Figure 3-2-4-1.

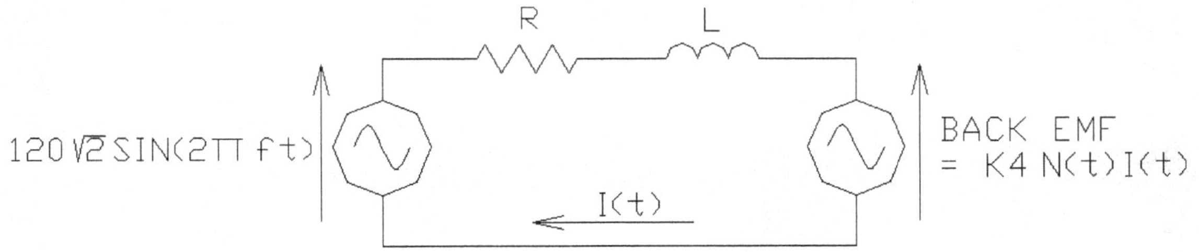

Figure 3-2-4-1 Time based motor equivalent circuit.

Mathcad's "solution of a system of differential equations". ability is demonstrated in this example.

Solution:

1) The equations needed are:

Motor output power (watts) PO(t) = I(t)^2·N(t)·K4

Motor output power (hp) POH(t) = PO(t)/746

Motor output torque (ft.lb.), T(t) = 5252·POH(t)/N(t)

Load and inertial torque (ft.lb.), $T(t) = K0 + K1 \cdot N(t) + K2 \cdot \sin[2 \cdot \pi \cdot N(t) \cdot f \cdot t/60] +$
$$K3 \cdot dN(t)/dt$$

Supply voltage (volts), $VS(t) = 120 \cdot 2^{.5} \cdot \sin(2 \cdot \pi \cdot f \cdot t)$

$$VS(t) = R \cdot I(t) + L \cdot dI(t)/dt + K4 \cdot N(t) \cdot I(t)$$

2) The equations need to be combined and manipulated so that the unknown differentials are on the left hand side of its "Ctrl=" sign and variables being solved are on the right side of its "Ctrl=" sign. The resulting equations and their initial conditions can be seen in the program in Figure 3-2-4-2.

Figure 3-2-4-2.xmcd SERIES MOTOR POWERING A RECIPROCATING PUMP, SYSTEM OF DIFFERENTIAL EQUATIONS

Defining fixed variables:

$R := 300$ $L := 1$ $f := 60$ $K4 := 10$

$K0 := .01$ $K1 := .00001$ $K2 := 1$ $K3 := .0001$

Given

$$N'(t) = \frac{1}{K3} \cdot \left(5252 \cdot \frac{I(t)^2 \cdot K4}{746} - K0 - K1 \cdot N(t) - K2 \cdot \sin\left(2 \cdot \pi \cdot \frac{N(t)}{60} \cdot t\right)\right)$$

<Note "Ctrl=" after N'(t), N(0),
                      I'(t), and I(0).>

$$N(0) = 0$$

$$I'(t) = \frac{1}{L} \cdot \left[120 \cdot (2)^{.5} \cdot \sin(2 \cdot \pi \cdot f \cdot t) - R \cdot I(t) - K4 \cdot N(t) \cdot I(t) \right]$$

$$I(0) = 0$$

Continued

$$\binom{I}{N} := \text{Odesolve}\left[\binom{I}{N}, t, 10\right]$$

$$T(t) := I(t)^2 \cdot K4 \cdot \frac{5252}{746}$$

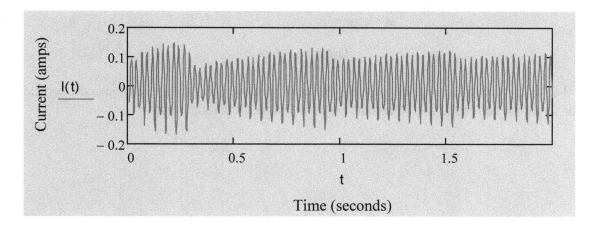

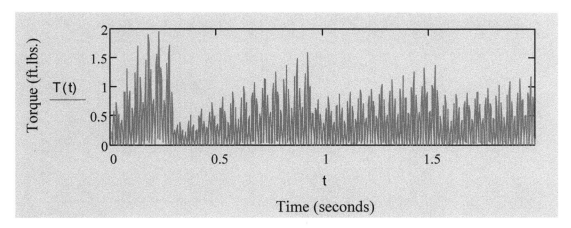

Continued

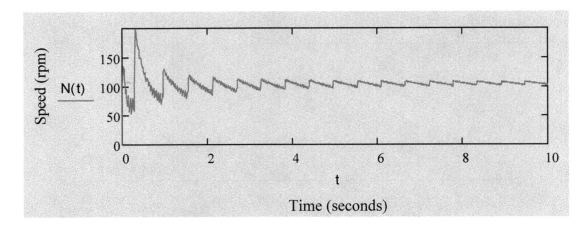

Figure 3-2-4-2 Mathcad program plotting the current waveform, torque, and speed of a series motor for .2 seconds after starting. The motor is driving a reciprocating pump.

3.3 MISCELLANEOUS EXAMPLES

3.3.1 DETERMINING THE AVERAGE, MEDIAN, AND STANDARD DEVIATION OF RESISTOR RESISTANCES

Problem:

Twenty-five resistors of the same 4700 Ω nominal value have their actual resistances measured. Using the actual resistance measurements, tabulate the resistance values, create a bar-chart histogram of them, and determine the resistance minimum, maximum, average, median, and standard deviation.

Mathcad's "statistics" and "bar-chart plotting" features are demonstrated in this example.

Solution:

1) Create the data table using the same method as in Section 3-1-9. Name this data table "Res" and make it 25 rows down and 1 column across.

2) Enter the resistor resistance values either by typing them in or by block copying them from a program like Microsoft Excel.

3) Type "Hi:histogram(8,Res)=. This defines "Hi" as a 2 by 8 matrix composed of eight equally spaced bins of resistance values.

4) From the main menu select "Insert", "Graph", and "X-Y Plot".

5) On the left center placeholder type "Hi<Ctrl>60". On the bottom center placeholder type "Hi<Ctrl>61". This will produce a point to straight line to point graph of the number of resistors in each bin of "Hi".

6) To create the bar chart, again select "Insert", "Graph" and "X-Y Plot". Double left-click on the plot. A "Formatting Currently Selected X-Y Plot" window appears. In this window left-click on "Traces" and change the "Type" for "Trace 1" from "Line" to "Bar". Then, left-click "OK".

7) To label the bar chart axes again double left-click on the plot. Left-click on the "Labels" tab. On the "X-Axis:" type *Resistor resistance (ohms)*. On the "Y-Axis:" type *Number of resistors*. Then, left-click "OK".

8) The median (most often occurring) resistor resistance is found with the "median(Res)" statement.

9) The average resistor resistance is found with the "mean(Res)" statement.

10) The standard deviation from the average resistor resistance is found with the "Stdev(Res)" statement.

70

Figure 3-3-1-1.xmcd STATISTICAL ANALYSIS OF RESISTOR RESISTANCES

Tabulated resistance values (ohms):

Res :=

	0
0	$4.7 \cdot 10^3$
1	$4.8 \cdot 10^3$
2	$4.9 \cdot 10^3$
3	$4.7 \cdot 10^3$
4	$4.5 \cdot 10^3$
5	$4.4 \cdot 10^3$
6	$5 \cdot 10^3$
7	$4.3 \cdot 10^3$
8	$5.1 \cdot 10^3$
9	$4.7 \cdot 10^3$
10	$4.8 \cdot 10^3$
11	$4.6 \cdot 10^3$
12	$4.9 \cdot 10^3$
13	$4.5 \cdot 10^3$
14	$5 \cdot 10^3$
15	$4.7 \cdot 10^3$
16	$4.6 \cdot 10^3$
17	$4.6 \cdot 10^3$
18	$4.6 \cdot 10^3$
19	$4.6 \cdot 10^3$
20	$4.5 \cdot 10^3$
21	$4.9 \cdot 10^3$
22	$4.4 \cdot 10^3$
23	$5 \cdot 10^3$
24	$4.7 \cdot 10^3$

Continued

Results:

$$Hi := \text{histogram}(8, \text{Res}) = \begin{pmatrix} 4.35 \times 10^3 & 1 \\ 4.45 \times 10^3 & 2 \\ 4.55 \times 10^3 & 3 \\ 4.65 \times 10^3 & 5 \\ 4.75 \times 10^3 & 7 \\ 4.85 \times 10^3 & 3 \\ 4.95 \times 10^3 & 3 \\ 5.05 \times 10^3 & 1 \end{pmatrix}$$

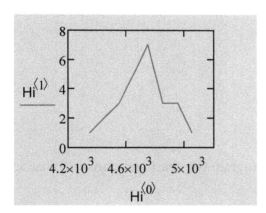

Continued

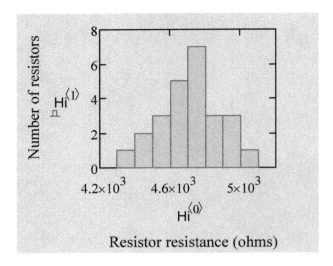

$$\text{min}(\text{Res}) = 4.3 \times 10^3$$

$$\text{max}(\text{Res}) = 5.1 \times 10^3$$

$$\text{median}(\text{Res}) = 4.7 \times 10^3$$

$$\text{mean}(\text{Res}) = 4.7 \times 10^3$$

$$\text{Stdev}(\text{Res}) = 212.132$$

Figure 3-3-1-1 Mathcad program producing a histogram bar chart and statistical characteristics of 25 resistors of varying resistance.

3.3.2 LEAST SQUARES FIT OF DIELECTRIC STRENGTH AND THICKNESS DATA TO A LINEAR EQUATION

Problem:

The dielectric breakdown strength (kV/mm) of electrical insulation is measured experimentally for different insulation thickness. The data does not indicate an obvious equation relating breakdown strength to thickness. What is the least-square best fit linear relationship between the insulation's dielectric breakdown strength and thickness?

Mathcad's "linear regression slope" and "linear regression intercept" features are demonstrated in this example.

Solution:

1) Create the data table using the same method as in Section 3-1-9. Name the data table "DES" and make it 9 rows down and 2 columns across. Either directly type in the data or block copy it in from an Excel spreadsheet.

2) Enter the program as shown in Figure 3-3-2-1. Make the plots using the methods of Section 3.1.7.

Figure 3-3-2-1.xmcd LEAST SQUARE REGRESSION ANALYSIS OF DIELECTRIC BREAKDOWN STRENGTH VERSUS THICKNESS DATA

DES :=

	0	1
0	1	10
1	0.9	8
2	0.8	12
3	0.6	15
4	2.1	3
5	1.4	5
6	2	2
7	0.9	11
8	0.8	9

Continued

74

$b0 := \text{intercept}\left(DES^{\langle 0 \rangle}, DES^{\langle 1 \rangle}\right) = 16.814$

$b1 := \text{slope}\left(DES^{\langle 0 \rangle}, DES^{\langle 1 \rangle}\right) = -7.269$

$DESeq := b0 + b1 \cdot DES^{\langle 0 \rangle}$

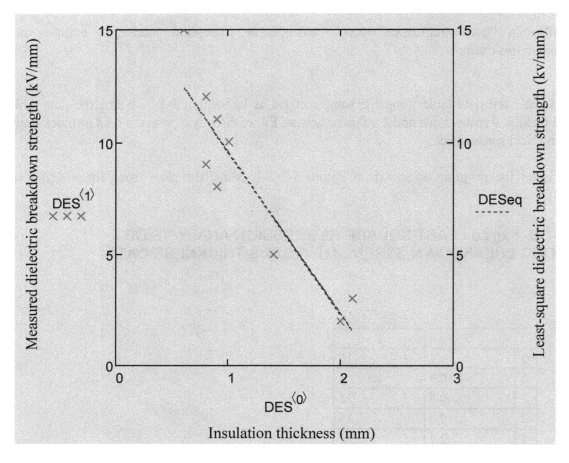

Figure 3-3-2-1 Mathcad program producing a plot and linear best-fit of dielectric breakdown versus thickness data.

3.3.3   APPROXIMATING RESISTANCE VERSUS TEMPERATURE DATA
WITH A POLYNOMIAL

Problem:

A 10 kΩ resistor has its resistance measured at different temperatures from -50° C to +150° C. Determine a best fit polynomial equation for its resistance versus temperature.

Mathcad's "minimization of least-square error, Minerr" feature is demonstrated in this example.

Solution:

1) In Mathcad create vectors representing the resistor resistances and their corresponding temperatures.

2) Write guess values for the polynomial coefficients. Sometimes, it is important for the guess values to be close to the correct final values to assure that the program does not converge on incorrect values. However, in this example the values are not too important.

3) Write a "Solve block" with the "Minerr" function.
     a) The "Solve block" is the part of the program from "Given" to "Minerr".
     b) Inside the "Solve block" write the equation to be solved by the "Minerr" function.
     c) The equation uses the "Ctrl=" that is used in all solve blocks.
     d) The values "R" and "T" are vectors with 20 values each. In the "Solve Block" the one "$R = a_0 + a_1 \cdot T + a_2 \cdot T^2 + a_3 \cdot T^3$" equation actually represents 20 equations.
     e) The coefficients a_0, a_1, a_2, and a_3 are array subscripted variables that are part of the "a" vector. The numbers refer to their position in the "a" vector. These subscripts are different from the visually identical "literal" subscripts that Mathcad can use. Literal subscripts are just another way of entering alphanumerics. To avoid confusion, "literal" subscripts are not used in this book.
          To make an "array subscripted variable" in Mathcad put the "[" symbol between the letter and integer number. For example to make a_0 type a[0.
     f) The "a := Minerr(a)" statement directs Mathcad to find "a" vector values that will make the selected polynomial have the least-square difference from the R data at each temperature.

4) Note that "Minerr" could also be used to solve non-polynomial equations.

5) The polynomial equation is evaluated at the specific data temperature points. The data and polynomial values are plotted to check the validity of the polynomial. As the plots show, the polynomial produces values very close to the original data.
     It is important to check the output of the "Minerr" function versus original data. In some cases "Minerr" will produce an incorrect solution.

6) See the program in Figure 3-3-3-1.

Figure 3-3-3-1.xmcd POLYNOMIAL REPRESENTING A RESISTOR'S
RESISTANCE VERSUS TEMPERATURE DATA

$$T := \begin{pmatrix} -50 \\ -40 \\ -30 \\ -20 \\ -10 \\ 0 \\ 10 \\ 20 \\ 30 \\ 40 \\ 50 \\ 60 \\ 70 \\ 80 \\ 90 \\ 100 \\ 110 \\ 120 \\ 130 \\ 140 \\ 150 \end{pmatrix} \qquad R := \begin{pmatrix} 10.32 \\ 10.27 \\ 10.2 \\ 10.13 \\ 10.09 \\ 10.04 \\ 10.02 \\ 10.01 \\ 10 \\ 10 \\ 10 \\ 10.01 \\ 10.03 \\ 10.05 \\ 10.09 \\ 10.14 \\ 10.2 \\ 10.29 \\ 10.36 \\ 10.44 \\ 10.55 \end{pmatrix}$$

Guess

$a_0 := 1$ $\qquad a_1 := 1$ $\qquad a_2 := 1$ $\qquad a_3 := 1$

Continued

Given

$$R = a_0 + a_1 \cdot T + a_2 \cdot T^2 + a_3 \cdot T^3$$ <Note "Ctrl=" after R>

$$a := Minerr(a)$$

$$a = \begin{pmatrix} 10.057 \\ -3.392 \times 10^{-3} \\ 4.06 \times 10^{-5} \\ 2.506 \times 10^{-8} \end{pmatrix}$$

$$Rpoly := a_0 + a_1 \cdot T + a_2 \cdot T^2 + a_3 \cdot T^3$$

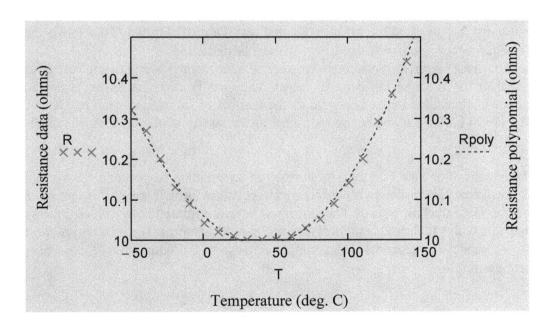

Figure 3-3-3-1 Mathcad program determining a polynomial representing a resistor's resistance versus temperature data.

3.3.4 DETERMINING THE FREQUENCY CONTENT OF A WAVEFORM WITH THE FAST FOURIER TRANSFORM

Mathcad's Fast Fourier Transform (FFT) functions determine the frequency spectrum of waveforms. They are particularly useful in the analysis of filters and in signal processing.

Mathcad is capable of two types of Fast Fourier Transforms and two types of Discrete Fourier Transforms. The common, scaled by 1/N, Fast Fourier Transform will be demonstrated here.

Characteristics of the Fast Fourier Transform:
1) It requires evenly spaced waveform samples.
2) To avoid aliasing errors, sampling must be at least twice the frequency of the highest waveform frequency. Generally, for good results, the sampling frequency should be at least six times the highest waveform frequency.
3) The number of samples must be exactly 2^n where n is some positive integer.
4) Results vary with the number of periods sampled. Generally, at least 3 periods should be sampled. Fractions of periods may be included in the sampling.
5) The Fast Fourier Transform is a special case of the Discrete Fourier Transform. The Discrete Fourier Transform will analyze waveforms containing any number of samples, not just those with 2^n samples.
6) The Fast Fourier Transform will produce a single frequency spectrum, waveform amplitude versus frequency. This is different that the Discrete Fourier Transform, that produces two frequency spectrums, one of which must be discarded.
7) The Fast Fourier Transform requires the computer to perform far fewer calculations than the Discrete Fourier Transform. The Fast Fourier Transform requires a number of calculations directly proportional to the number of sample points. The Discrete Fourier Transform requires a number proportional to the square of the number of sampled points.

Many demonstrations of Fast Fourier Transforms use an equation to generate a waveform for analysis. However, actual electrical data is not received in equation form. It appears in a graphical or numerical data form. Here the Fast Fourier Transform will be demonstrated both with an equation generated waveform and with a tabulated numerical data generated waveform. To make comparisons easier, the data of the tabulated numerical data example came from the equation of the equation generated waveform example.

3.3.4.1 Equation Generated Waveform Example of the Fast Fourier Transform

Problem:

Determine the Fast Fourier Transform frequency spectrum for the waveform described by the equation:

$$f(t) = \sin(w_0 t) + 2 \cdot \sin(2 \cdot w_0 \cdot t + .7854) + 3 \cdot \sin(3 \cdot w_0 \cdot t)$$

Mathcad's "Fast Fourier Transform" is demonstrated in this example.

Solution:

1) Enter the program shown in Figure 3-3-4-1-1.

Figure 3-3-4-1-1.xcmd FAST FOURIER TRANSFORM OF EQUATION GENERATED WAVEFORM DATA

$w_0 := 377$ Angular frequency of the waveform being analyzed, radians/second

$f(t) := \sin(w_0 \cdot t) + 2\sin(2 \cdot w_0 \cdot t + .7854) + 3 \cdot \sin(3 \cdot w_0 \cdot t)$ Waveform equation

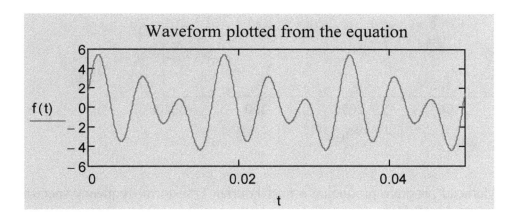

Continued

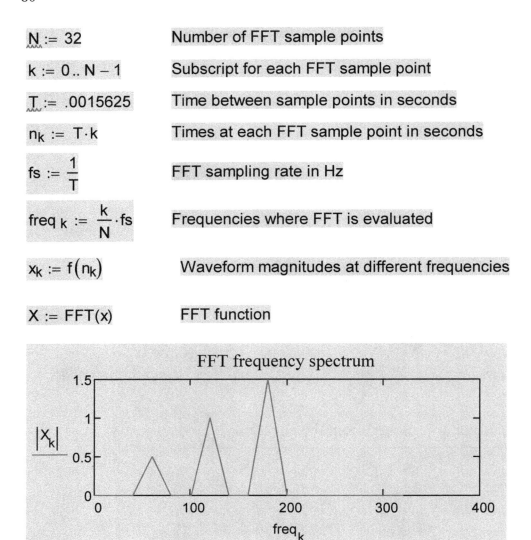

$N := 32$ — Number of FFT sample points

$k := 0 .. N - 1$ — Subscript for each FFT sample point

$T := .0015625$ — Time between sample points in seconds

$n_k := T \cdot k$ — Times at each FFT sample point in seconds

$fs := \dfrac{1}{T}$ — FFT sampling rate in Hz

$freq_k := \dfrac{k}{N} \cdot fs$ — Frequencies where FFT is evaluated

$x_k := f(n_k)$ — Waveform magnitudes at different frequencies

$X := FFT(x)$ — FFT function

Figure 3-3-4-1-1 Mathcad program producing a Fast Fourier Transform frequency spectrum from an equation-generated waveform.

2) The created function is plotted to check that it looks reasonable.

3) After the plot there are a series of program lines that carry out the Fast Fourier Transform. Descriptions of these are written as comments into the program itself. See Figure 3-3-4-1-1.

4) Following the Fast Fourier Transform lines, an ordinary 2-D plot is made of magnitude versus frequency. Note that the peak frequencies in Hz in the plot are the same as the frequencies in the original equation of the waveform.

3.3.4.2 Tabulated Data Generated Waveform Example of the Fast Fourier Transform

Problem:

Use Mathcad's Fast Fourier Transform function to determine the frequency spectrum for the waveform described in the following data.

Mathcad's "Data Import Wizard" is demonstrated in this example.

Solution:

1) Create a data file for the waveform amplitude versus time. As in Section 3.3.4.1 the time should vary from 0 to .049 seconds. These numbers could by typed directly into a Mathcad data chart. However an easier way is to use the Microsoft Excel program to generate the numbers, and then import the created data chart. Figure 3-3-4-2-1 shows the numbers on an Excel spreadsheet.

2) In the Excel program, formulas can be used to generate the data chart. Make column A row 0 equal to 0. After that, in column A row 1 enter the formula "=.001+A1". Then, copy that formula to the blocks below it. The formula will automatically change A1 to A2, etc. and fill in the times.   Likewise,   in   column   B   row   1   enter   the   formula "=SIN(377*A1)+2*SIN(2*377*A1+0.7854)+3*SIN(3*377*A1)". Then, copy this to the locations below it to fill in the amplitudes. As mentioned earlier, the amplitude data does not have to follow the "=SIN(377*A1….)" equation, or any other equation. The "=SIN(377*A1….)" equation is used here to simplify the generation of three periods of a waveform and to make it easier to compare with the ideal case where the periodic waveform can be described by a simple periodic equation.

0	1.414
1.00E-03	5.082
2.00E-03	4.496
3.00E-03	0.347
4.00E-03	-3.175
5.00E-03	-2.787
6.00E-03	0.562
7.00E-03	3.040
8.00E-03	2.247
9.00E-03	-0.382
1.00E-02	-1.659
1.10E-02	-0.542

Continued

1.20E-02	0.757
1.30E-02	-0.286
1.40E-02	-3.103
1.50E-02	-4.349
1.60E-02	-1.743
1.70E-02	2.953
1.80E-02	5.428
1.90E-02	3.338
2.00E-02	-1.127
2.10E-02	-3.529
2.20E-02	-1.831
2.30E-02	1.675
2.40E-02	3.145
2.50E-02	1.413
2.60E-02	-1.085
2.70E-02	-1.495
2.80E-02	0.031
2.90E-02	0.723
3.00E-02	-1.165
3.10E-02	-3.871
3.20E-02	-3.911
3.30E-02	-0.208
3.40E-02	4.226
3.50E-02	5.222
3.60E-02	1.894
3.70E-02	-2.348
3.80E-02	-3.385
3.90E-02	-0.658
4.00E-02	2.535
4.10E-02	2.859
4.20E-02	0.489
4.30E-02	-1.525
4.40E-02	-1.090
4.50E-02	0.501
4.60E-02	0.367
4.70E-02	-2.153
4.80E-02	-4.320
4.90E-02	-3.016

Figure 3-3-4-2-1 Tabulated waveform time (column A) and amplitude (column B) on a Microsoft Excel spreadsheet for use on a Fast Fourier Transform.

3) Create the Mathcad program shown in Figure 3-3-4-2-2.

4) To put the Excel data file into a Mathcad data chart with the "Data Import Wizard":
    a) Position the Mathcad cursor at the desired location.
    b) Left-click "Insert", "Data", and "Data Import Wizard".
    c) Left-click "Microsoft Excel" as the "File Format".
    d) Left-click "Browse" and "Next".
    e) Left-click and "Open" the appropriate file.
    f) Left-click on "Finish".
    g) A data chart with column 0 to 3 and rows 0 to 15 appears. Columns 2 and 3 have "NaN", indicating that they are Not a Number. Mathcad ignores these columns.
    h) If desired, the data sheet could be stretched down to show all of the data.
    i) Type *Data* over the placeholder at the top left of the chart.

5) The tabulated data is now converted to a function that approximates it with linear interpolation. The created function is plotted to check that it looks reasonable. Compare the linear interpolation plot of $f(t)$ in Figure 3-3-4-2-2 with that of the original function $f(t)$ in Figure 3-3-4-1-1.

84

Figure 3-3-4-2-2.xmcd FAST FOURIER TRANSFORMATION OF
TABULATED WAVEFORM DATA

Data :=

	0	1	2	3
0	0	1.414	NaN	NaN
1	$1 \cdot 10^{-3}$	5.082	NaN	NaN
2	$2 \cdot 10^{-3}$	4.496	NaN	NaN
3	$3 \cdot 10^{-3}$	0.347	NaN	NaN
4	$4 \cdot 10^{-3}$	-3.175	NaN	NaN
5	$5 \cdot 10^{-3}$	-2.787	NaN	NaN
6	$6 \cdot 10^{-3}$	0.562	NaN	NaN
7	$7 \cdot 10^{-3}$	3.04	NaN	NaN
8	$8 \cdot 10^{-3}$	2.247	NaN	NaN
9	$9 \cdot 10^{-3}$	-0.382	NaN	NaN
10	0.01	-1.659	NaN	NaN
11	0.011	-0.542	NaN	NaN
12	0.012	0.757	NaN	NaN
13	0.013	-0.286	NaN	NaN
14	0.014	-3.103	NaN	NaN
15	0.015	-4.349	NaN	...

$f(t) := \text{linterp}\left(\text{Data}^{\langle 0 \rangle}, \text{Data}^{\langle 1 \rangle}, t\right)$ Function that makes a linear
interpolation of the data

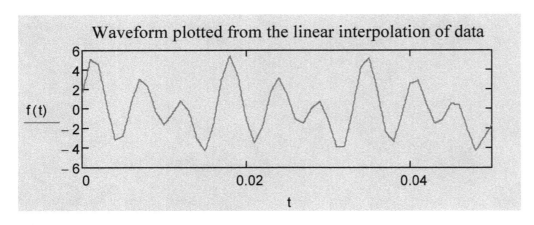

Waveform plotted from the linear interpolation of data

Continued

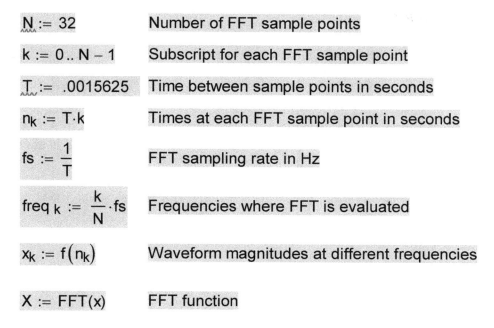

$N := 32$ Number of FFT sample points

$k := 0 .. N - 1$ Subscript for each FFT sample point

$T := .0015625$ Time between sample points in seconds

$n_k := T \cdot k$ Times at each FFT sample point in seconds

$fs := \dfrac{1}{T}$ FFT sampling rate in Hz

$freq_k := \dfrac{k}{N} \cdot fs$ Frequencies where FFT is evaluated

$x_k := f(n_k)$ Waveform magnitudes at different frequencies

$X := FFT(x)$ FFT function

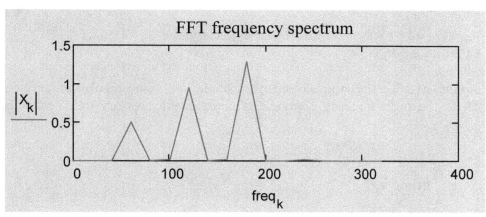

Figure 3-3-4-2-2 Mathcad program producing a Fast Fourier Transform frequency spectrum from a tabulated-data generated waveform.

6) The same Fast Fourier Transform commands as were used in Section 3.3.4.1 are used here to produce a frequency spectrum plot. Compare the frequency spectrum plot of Figure 3-3-4-2-2 with that of the original function in Figure 3-3-4-1-1. Note that the frequency plot of Figure 3-3-4-2-2 has the same frequency peaks as the frequency spectrum plot of Figure 3-3-4-1-1, although its magnitudes are less.

3.3.5 FOURIER SERIES ANALYSIS

A Fourier Series is a representation of a periodic waveform by a sum of sinusoids. This is different from a Fourier Transform which produces an amplitude versus frequency spectrum

Unlike the Fast Fourier Transform, the finding of Fourier Coefficients requires computations be done over an integral number of periods. Usually one period is selected. Also, unlike the Fast Fourier Transform, the number of samples does not have to be exactly 2^n (n is an integer) and the samples do not need to be evenly spaced.

With a Fourier Series a periodic waveform, f(t), can be represented as a sum of sine and cosine waves or the sum of sine waves with differing phase angles. The waves have frequencies that are multiples of a primary frequency. The Fourier Series representation of a wave is:

$$f(t) = A_0 + A_1 \cdot \sin(w \cdot t) + A_2 \cdot \sin(2 \cdot w \cdot t) + A_3 \cdot \sin(3 \cdot w \cdot t) + \ldots + A_n \cdot \sin(n \cdot w \cdot t) +$$
$$B_1 \cdot \cos(w \cdot t) + B_2 \cdot \cos(2 \cdot w \cdot t) + B_3 \cdot \cos(3 \cdot w \cdot t) + \ldots + B_n \cdot \cos(n \cdot w \cdot t)$$

or

$$f(t) = A_0 + C_1 \cdot \sin(w \cdot t + \theta_1) + C_2 \cdot \sin(2 \cdot w \cdot t + \theta_2) + C_3 \cdot \sin(3 \cdot w \cdot t + \theta_3) + \ldots +$$
$$C_n \cdot \sin(n \cdot w \cdot t + \theta_n)$$

The function of time, f(t), describes the waveform amplitude, "w" is the angular frequency of the base harmonic, "n" is the order of each harmonic. Fourier coefficients are found with the following equations:

$$A_0 = (1/T) \cdot \int_0^T f(t) \, dt$$

$$A_n = (2/T) \cdot \int_0^T f(t) \cdot \sin(n \cdot w_0 \cdot t) \, dt$$

$$B_n = (2/T) \cdot \int_0^T f(t) \cdot \cos(n \cdot w_0 \cdot t) \, dt$$

$$C_n = |A_n + B_n \cdot j|$$

$$\theta_n = \arg(A_n + B_n \cdot j)$$

T is the period of the base (lowest frequency) harmonic.

Problem:

Determine the Fourier Series equation that represents the waveform of the data of Section 3.3.4.2. As in Section 3.3.4.2, the base harmonic angular frequency is 377 radians/second.

Mathcad's "definite integral" feature is demonstrated in this example.

Solution:

1) Make an Excel file for one period of the waveform. The base harmonic angular frequency of 377 radians/second produces a period of $2\pi/377$ seconds. The data in Figure 3-3-4-2-1 can be used, except for the last time. Change the last time from .017 to $2\pi/377$. Using the "=SIN(377*A1….)" equation that was used in Section 3.3.4.2, recalculate the amplitude at that time. The data from the Excel spreadsheet is in Figure 3-3-5-1.

0	1.414
1.00E-03	5.082
2.00E-03	4.496
3.00E-03	0.347
4.00E-03	-3.175
5.00E-03	-2.787
6.00E-03	0.562
7.00E-03	3.040
8.00E-03	2.247
9.00E-03	-0.382
1.00E-02	-1.659
1.10E-02	-0.542
1.20E-02	0.757
1.30E-02	-0.286
1.40E-02	-3.103
1.50E-02	-4.349
1.60E-02	-1.743
1.67E-02	1.416

Figure 3-3-5-1 Tabulated waveform time (column A) and amplitude (column B) on a Microsoft Excel spreadsheet for one period of the waveform.

2) As with the Fourier Transform, the Fourier Series does not need to be done with an equation like "=SIN(377*A1….)" or any other equation. It can be done with data without the involving of any equation. The "=SIN(377*A1….)" equation is only used here for the convenience of being able to easily generate data points.

3) Create the Mathcad program shown in Figure 3-3-5-2.

88

Figure 3-3-5-2.xmcd FOURIER SERIES OF A WAVEFORM DESCRIBED BY
TABULATED DATA

$w_0 := 377$ Angular frequency
(radians/second)

$T := 2 \cdot \dfrac{\pi}{w_0}$ Period of waveform
(seconds)

Data :=

	0	1
0	0	1.414
1	$1 \cdot 10^{-3}$	5.082
2	$2 \cdot 10^{-3}$	4.496
3	$3 \cdot 10^{-3}$	0.347
4	$4 \cdot 10^{-3}$	-3.175
5	$5 \cdot 10^{-3}$	-2.787
6	$6 \cdot 10^{-3}$	0.562
7	$7 \cdot 10^{-3}$	3.04
8	$8 \cdot 10^{-3}$	2.247
9	$9 \cdot 10^{-3}$	-0.382
10	0.01	-1.659
11	0.011	-0.542
12	0.012	0.757
13	0.013	-0.286
14	0.014	-3.103
15	0.015	-4.349
16	0.016	-1.743
17	0.017	1.416

$f(t) := \text{linterp}\left(\text{Data}^{\langle 0 \rangle}, \text{Data}^{\langle 1 \rangle}, t\right)$

Function that does the linear
interpolation of the data

Continued

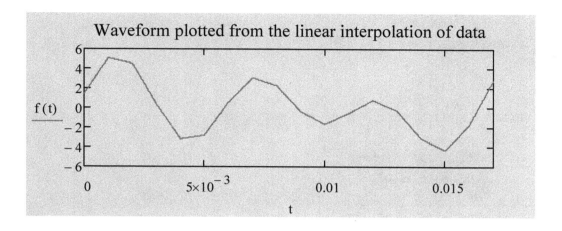

Waveform plotted from the linear interpolation of data

Evaluation of Fourier coefficients

$$A_0 := \frac{1}{T} \cdot \int_0^T f(t)\, dt = -4.566 \times 10^{-3}$$

$$A_1 := \frac{2}{T} \cdot \int_0^T f(t) \cdot \sin(w_0 \cdot t)\, dt = 0.989$$

$$A_2 := \frac{2}{T} \cdot \int_0^T f(t) \cdot \sin(2\, w_0 \cdot t)\, dt = 1.35$$

$$A_3 := \frac{2}{T} \cdot \int_0^T f(t) \cdot \sin(3\, w_0 \cdot t)\, dt = 2.695$$

$$A_4 := \frac{2}{T} \cdot \int_0^T f(t) \cdot \sin(4\, w_0 \cdot t)\, dt = 1.895 \times 10^{-3}$$

$$A_5 := \frac{2}{T} \cdot \int_0^T f(t) \cdot \sin(5\, w_0 \cdot t)\, dt = 2.578 \times 10^{-3}$$

Continued

$$B_1 := \frac{2}{T} \cdot \int_0^T f(t) \cdot \cos(w_0 \cdot t) \, dt = -9.297 \times 10^{-3}$$

$$B_2 := \frac{2}{T} \cdot \int_0^T f(t) \cdot \cos(2 w_0 \cdot t) \, dt = 1.339$$

$$B_3 := \frac{2}{T} \cdot \int_0^T f(t) \cdot \cos(3 w_0 \cdot t) \, dt = -9.102 \times 10^{-3}$$

$$B_4 := \frac{2}{T} \cdot \int_0^T f(t) \cdot \cos(4 w_0 \cdot t) \, dt = -0.011$$

$$B_5 := \frac{2}{T} \cdot \int_0^T f(t) \cdot \cos(5 w_0 \cdot t) \, dt = -9.399 \times 10^{-3}$$

Fourier coefficients expressed as a magnitude and angle

$$C_1 := |A_1 + B_1 \cdot j| = 0.989$$
$$\theta_1 := \arg(A_1 + B_1 \cdot j) = -9.402 \times 10^{-3}$$

$$C_2 := |A_2 + B_2 \cdot j| = 1.901$$
$$\theta_2 := \arg(A_2 + B_2 \cdot j) = 0.782$$

$$C_3 := |A_3 + B_3 \cdot j| = 2.695$$
$$\theta_3 := \arg(A_3 + B_3 \cdot j) = -3.378 \times 10^{-3}$$

Continued

$$C_4 := \left| A_4 + B_4 \cdot j \right| = 0.011$$

$$\theta_4 := \arg\left(A_4 + B_4 \cdot j \right) = -1.398$$

$$C_5 := \left| A_5 + B_5 \cdot j \right| = 9.746 \times 10^{-3}$$

$$\theta_5 := \arg\left(A_5 + B_5 \cdot j \right) = -1.303$$

Resultant Fourier Series, ignoring small coefficients and angles.

$$fr(t) := A_0 + C_1 \cdot \sin\left(w_0 \cdot t\right) + C_2 \sin\left(2 \cdot w_0 \cdot t + \theta_2\right) + C_3 \cdot \sin\left(3 \cdot w_0 \cdot t\right)$$

Fourier Series equation waveform drawn over the waveform of the original data

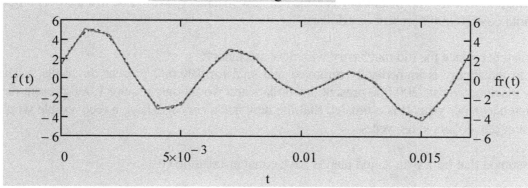

Figure 3-3-5-2 Mathcad program of a Fourier Series analysis of a waveform described by tabulated data. f(t) is the linearly interpolated plot of the original data. fr(t) is the plot of the Fourier Series equation.

3.3.6 ENGINEERING ECONOMICS STUDY

Engineering economics includes the calculation of present values, payment schedules, and future values. Engineers need to be able to do these calculations to help their companies be profitable.

Problem:
A company is considering replacing old machinery with equivalent capacity and efficiency new machinery by either plan A or plan B. Use Mathcad to determine the values of plan A and B at ten years.

Plan A-Keep the old machinery for ten more years and then replace it:
The old machinery requires 3,000 present day dollars of maintenance per year. It is estimated that it will only function for 10 more years, after that it will be removed, sold, and scrapped for 5,000 present day dollars. After the old machinery is removed new machinery would be installed. The new machinery would cost 200,000 present day dollars.

Plan B-Immediately replace the old machinery with new machinery:
The old machinery is immediately removed and sold for 100,000 present day dollars. The proposed new machinery costs 200,000 present day dollars and would only require 1,000 present day dollars of maintenance per year. It is estimated that the new machinery will have a recoverable worth of 90,000 present day dollars in ten years.

It is assumed that both plan A and plan B have equal installation costs.

It is estimated that interest due on borrowed money will be 7% and inflation will be 4%.

Once a machine is installed its resale value drops by 15%.

In each plan money is borrowed.

Maintenance and payments will be made at the end of each year.

Mathcad's "fv (future value)" and "pmt (payment)" functions are demonstrated in this example.

Solution:
1) For each plan, the future value costs will be calculated and then subtracted from the future value's estimated asset value.

2) Write the program of Figure 3-3-6-1. Comments in the program describe the calculations.

Figure 3-3-6-1.xmcd ENGINEERING ECONOMIC COMPARISON STUDY

Borrowing interest rate	$BIR := .07$
Inflation rate	$IR := .04$

PLAN A

ITEM	VALUE ($)

Estimated new machinery purchase price at ten years accounting for inflation. Money paid at ten years.

$P1FVA := (200000) \cdot (1 + IR)^{10}$

$P1FVA = 2.96 \times 10^5$

Projected cost of ten annual old machinery $3000 maintenance payments after ten years. Annual payments are made at the end of each year.

$P2FVA := fv(BIR, 10, -3000, 0, 0)$

$P2FVA = 4.145 \times 10^4$

Estimated recoverable worth of new machinery after ten years.

$RNFVA := P1FVA \cdot (1 - .15) = 2.516 \times 10^5$

Estimated recoverable worth of old machinery after ten years.

$ROFVA := (5000) \cdot (1 + IR)^{10}$

$ROFVA = 7.401 \times 10^3$

RESULT: Recoverable worth minus total money paid out at ten years

$RNFVA + ROFVA - P1FVA - P2FVA = -7.846 \times 10^4$

Continued

PLAN B

ITEM	VALUE ($)

Old machinery selling price (sold at beginning of ten years).

$ROPVB := 100000$

New machinery cost minus old machinery selling price at beginning of ten years.

$P1PVB := 200000 - ROPVB = 1 \times 10^5$

New machinery minus old machinery projected cost after 10 years.

$P1FVB := P1PVB \cdot (1 + BIR)^{10}$

$P1FVB = 1.967 \times 10^5$

Projected cost of ten annual new machinery $1000 maintenance payments after ten years. Annual payments are made at the end of each year.

$P2FVB := fv(BIR, 10, -1000, 0, 0)$

$P2FVB = 1.382 \times 10^4$

Estimated recoverable worth of new machinery after ten years.

$RFVB := 90000 \cdot (1 + IR)^{10}$

$RFVB = 1.332 \times 10^5$

RESULT: Recoverable worth minus total money paid out at ten years

$RFVB - P1FVB - P2FVB = -7.731 \times 10^4$

Figure 3-3-6-1 Mathcad economic comparison study of possible machinery replacement plans.

3) The cost of plan B is very close to that of plan A. If the rates and prices were to change, that relationship could change. Once the Mathcad program is written, prices and rates can easily be changed to see their effect on future values.

4.0 MATHCAD IN REPORTS

Mathcad programs are easier to read than many other analysis programs. Simpler Mathcad programs can be understood even by those who have never used it. Therefore reports using Mathcad programs tend to be clearer.

Mathcad will function as a word processor, handling both text and calculations. It will also allow the importing of graphics. However, it takes extra effort to produce a Mathcad document that is as neat as one made with a word processor. Mathcad does not have the text processing features of word processing programs. For most reports, it is better to word process the text and block copy and insert the Mathcad program into the word processed report.

If a report is mostly a Mathcad program with little text, it would be reasonable to use Mathcad for the whole report. Mathcad would also be a more reasonable choice if a report's input data or equations are often changed.

5.0 REFERENCES

1) *User's Guide*, Mathcad 14.0, Parametric Technology Corp., 2007. It is included with the Mathcad CD. Free copies of it have also been found on the internet.
It prints on 166 pages. Anybody using Mathcad should get a copy.

2) *Electrical and Electronic Engineering*, E-book adaptation of the electrical engineering and electronics engineering section, pages 4.1 to 5.80, of the *Standard Handbook of Engineering Calculations* edited by Tyler G. Hicks, Mathsoft Engineering & Education, Inc., 1997. If it was purchased with Mathcad, it can be found under Mathcad's "Help" heading, under "E-books".
It contains 72 sections, each devoted to a different electrical or electronic subject. Most of the sections contain useful equations. *Electrical and Electronic Engineering* is a good source of electrical and electronic knowledge, like the book it was copied from. However, it gives few details on the use of Mathcad.

3) *Electrical Power Systems Engineering*, E-book, Spezia, Carl J. and Hatziadoniu, Constantine I., Mathsoft Engineering & Education, Inc., 2004. If it was purchased with Mathcad, it can be found under Mathcad's "Help" heading, under "E-books".
It contains both electrical theory and application programs dealing with power distribution problems, protection issues, and electrical transient phenomena. It would be useful to one who wants to use Mathcad to solve electrical power systems problems.

4) *Topics in Electrical Engineering*, E-book, Mathsoft Engineering & Education, Inc., 2004. If it was purchased with Mathcad, it can be found under Mathcad's "Help" heading, under "E-books".
It contains application programs in electromagnetics, transmission lines/smith charts, circuit and feedback analysis, signal processing techniques, and filter and transfer functions. The electrical applications are sort that would be seen in electrical engineering graduate schools. According to its "About this E-book" section, "These applications have been developed as general-purpose tools that illustrate a variety of useful Mathcad techniques." *Topics in Electrical Engineering* would be useful to one doing high-level electrical research in its topics, but would not be of much use to others.

5) Maxfield, Brent, *Essential Mathcad for Engineering, Science, and Math w/CD*, second edition, 2009, Academic Press. It is paperback, 501 pages long, and 9.3" x 7.3". Its list price is $49.95. Maxfield's book brings in examples from a variety of fields. A program CD that comes with the book contains a full non-expiring, for educational purposes only, for North America only, student version of Mathcad. The inclusion of the program CD makes Maxfield's book a very good deal for academics.

6.0 APPENDIX

6.1 NUMERIC PRECISION

Internally Mathcad maintains all numbers to double precision (64 bit) floating point format. This means that decimal numbers are internally stored to at least 17 significant figures. Generally Mathcad rounds numbers that have more than 12 figures to the right of the decimal point.

By default, Mathcad displays 3 significant figures to the right of the decimal point. The displayed figures can be adjusted using the "Result Format" window in the "Format" menu.

6.2 TIPS

Double-check Mathcad results. Occasionally Mathcad will produce an incorrect result. This is especially possible when using the "Minerr" operator. With this Mathcad may stop evaluating before it has reached the correct solution.

Recalculate values in a worksheet after numbers in it have been changed. Sometimes Mathcad will not automatically recalculate values. This was seen in the example in Section 3.1.11. To make certain that Mathcad recalculates values use the "Calculate Worksheet" command. This command is in the "Tools" menu under "Calculate".

Use the "Ctrl=" where appropriate. In Mathcad the "Ctrl=" appears as a bold "=". Its appearance is so similar to an ordinary "=" that time may be wasted debugging programs where an incorrect equals sign was used.

6.3 MATHCAD OPERATORS

Many Mathcad operators can be entered in more than one way. Entry can be by selection from a toolbar, selection from a pull-down menu, or typing in from the keyboard. For example, the sine function can be selected from the arithmetic toolbar or typed into an equation with the keyboard.

The Mathcad "Insert Function" or "f(x)" can be accessed through the Standard Toolbar or pull-down menu. Its purpose is to insert properly formatted functions into Mathcad programs. However, for the person learning Mathcad its greatest value may be that it displays all the Mathcad functions in alphabetical order. Scrolling through the functions in the "Insert Function" window shows all of the available Mathcad functions. By left-clicking on a function a brief description of it appears.

As mentioned earlier, Mathcad has more operators than most electrical engineers need. Below is a listing of the families of operators demonstrated in this book.

Arithmetic Calculations
    Enter via the Calculator Toolbar and keyboard.
        See examples all sections.

Calculus
    Enter via the Calculus Toolbar and keyboard.
        Differential Equations
            Numerical solutions
                See examples in Sections 3.2.1, 3.2.2, & 3.2.4.
            Laplace Transforms
                See example in Section 3.2.3.
        Differentiation
            See example in Section 3.2.3,
        Integration
            See examples in Sections 3.1.11 & 3.3.5.

Best Fitting an Equation to Data
    Enter via keyboard.
        See examples in Sections 3.3.2, 3.3.3, & 3.3.5.

Complex Number Calculations
    Enter via the Calculator Toolbar and keyboard.
        See examples in Sections 3.1.3, 3.1.4, & 3.1.5

Data Tables
    Enter via the Standard Toolbar and pull-down menu.
        See examples in Sections 3.1.9, 3.1.10, 3.3.1, 3.3.2, 3.3.4.2, & 3.3.5.

Evaluation
>Enter via the keyboard, Evaluation toolbar, Calculator toolbar, Standard toolbar & Symbolic toolbar.
>>See examples in all sections.

Finance
>Enter via the keyboard.
>>See example in Section 3.3.6.

Matrices & Vectors
>Enter via the Matrix Toolbar and pull-down menu.
>>See examples in Sections 3.1.2.1, 3.1.2.2, 3.1.2.3, 3.1.5, 3.3.1, & 3.3.3.

Plotting and Graphing
>Enter via the Graph Toolbar and pull-down menu.
>>See examples in Sections 3.1.7, 3.1.8, 3.1.9, 3.1.10, 3.1.11, 3.2.1, 3.2.2, 3.2.3, 3.2.4, 3.3.1, 3.3.2, 3.3.3, 3.3.4, 3.3.5.

Programming (including Boolean Logic)
>Enter via the Programming Toolbar and Boolean Toolbar.
>>See example in Section 3.1.11.

Statistics
>Enter via the keyboard, and Standard Toolbar Insert Function.
>>See example in Section 3.3.1.

Symbolic Calculations
>Enter via the Symbolic Toolbar.
>>See examples in Sections 3.1.4 & 3.2.3.

Trigonometric Calculations
>Enter via the Calculator Toolbar and keyboard.
>>See examples in Sections 3.1.10, 3.2.4, 3.3.4, & 3.3.5.

Units
>Enter via the Standard Toolbar, keyboard, and pull-down menu.
>>See examples in Sections 3.1.1, 3.1.2.2, 3.1.2.3, 3.1.3, 3.1.6, & 3.1.7.

6.4 SYSTEM REQUIREMENTS FOR MATHCAD 14.0

Hardware Requirements

Pentium/Celeron-compatible 32-bit (x86) or 64-bit (x86-64, EM64T) processor 400 MHz or higher; 700+ MHz recommended.

256 MB of RAM; 512 MB or more recommended.

550 MB of hard disk space (250 MB for Mathcad, 100 MB for prerequisites, and 200 MB for temporary space during installation).

CD or DVD drive (for CD installation only).

SVGA or higher graphics card and monitor.

Mouse or compatible pointing device.

Software Requirements

Windows 2000 SP4, Windows XP Home or Professional Edition, Windows XP Professional x64 Edition, Windows 2003, or Windows Vista.

Microsoft .NET Framework® 2.0

MSXML 4.0 SP2 or later

Microsoft Data Access Components 2.6 or later

Internet Explorer 6.0 or later.

Adobe Reader 5.0 or later.